Discovery Travel Adventures™

W9-BNR-708

CAVES, CLIFFS
& CANYONS

Robert Burnham
Editor

John Gattuso
Series Editor

Discovery Communications, Inc.

Discovery Communications, Inc.
John S. Hendricks, *Founder, Chairman, and Chief Executive Officer*
Judith A. McHale, *President and Chief Operating Officer*
Michela English, *President, Discovery Enterprises Worldwide*
Judy L. Harris, *Senior Vice President, Consumer Products, Discovery Enterprises Worldwide*

Discovery Publishing
Natalie Chapman, *Vice President, Publishing*
Rita Thievon Mullin, *Editorial Director*
Mary Kalamaras, *Senior Editor*
Maria Mihalik Higgins, *Editor*
Heather Quinlan, *Editorial Coordinator*
Chris Alvarez, *Business Development*
Jill Gordon, *Marketing Specialist*

Discovery Channel Retail
Tracy Fortini, *Product Development*
Steve Manning, *Naturalist*

Insight Guides
Jeremy Westwood, *Managing Director*
Brian Bell, *Editorial Director*
John Gattuso, *Series Editor*
Siu-Li Low, *General Manager, Books*

Distribution
United States
Langenscheidt Publishers, Inc.
46-35 54th Road
Maspeth, NY 11378
Fax: 718-784-0640

Worldwide
APA Publications GmbH & Co. Verlag KG
Singapore Branch, Singapore
38 Joo Koon Road
Singapore 628990
Tel: 65-865-1600
Fax: 65-861-6438

© **2000** Discovery Communications, Inc., and Apa Publications GmbH & Co Verlag KG, Singapore Branch, Singapore. All rights reserved under international and Pan-American Copyright Conventions.

Discovery Travel Adventures™ and Explore a Passion, Not Just a Place™ are trademarks of Discovery Communications, Inc.

Discovery Communications produces high-quality nonfiction television programming, interactive media, books, films, and consumer products. Discovery Networks, a division of Discovery Communications, Inc., operates and manages the Discovery Channel, TLC, Animal Planet, Travel Channel, and Discovery Health Channel. Visit Discovery Channel Online at www.discovery.com.

Although every effort is made to provide accurate information in this publication, we would appreciate readers calling our attention to any errors or outdated information by writing us at: Insight Guides, PO Box 7910, London SE1 1WE, England; fax: 44-20-7403-0290;
e-mail: insight@apaguide.demon.co.uk

No part of this publication may be reproduced in any form or by any electronic or mechanical means, including information storage and retrieval devices or systems, without prior written permission from the publisher, except that brief passages may be quoted for reviews.

Printed by Insight Print Services (Pte) Ltd, 38 Joo Koon Road, Singapore 628990.

Cataloging-in-Publication Data for this book is on file with the Library of Congress, Washington, D.C.

*C*aves, Cliffs & Canyons combines the interests and enthusiasms of two of the world's best-known information providers: **Insight Guides**, whose titles have set the standard for visual travel guides since 1970, and **Discovery Communications**, the world's premier source of non-fiction entertainment. The editors of Insight Guides provide both practical advice and general understanding about a destination's history, culture, institutions, and people. Discovery Communications and its website, www.discovery.com, help millions of viewers explore their world from the comfort of their home and encourage them to explore it firsthand.

About This Book

This book reflects the contributions of dedicated editors and writers familiar with the most interesting sites in North American geology. Series editor **John Gattuso**, of Stone Creek Publications in New Jersey, worked with Insight Guides and Discovery Communications to conceive and direct the series. Gattuso looked to **Robert Burnham**, a Wisconsin-based science writer and author of several books, to serve as project editor. A founding editor of *Earth* magazine and former editor-in-chief of *Astronomy* magazine, Burnham began writing about our planet and its neighboring worlds nearly 25 years ago. "I grew up during the Space Age," says Burnham, "and I've always been fascinated by the Moon and planets. My curiosity to know about Earth and how it works grew out of that interest. After all, it's the only planet we can explore firsthand." In addition to editing the book, Burnham wrote about Meteor Crater in northern Arizona: "The most common landforms in the solar system aren't mountains, tectonic faults, or volcanos, but craters caused by the impact of meteorites and comets."

To cover the background chapters on geological science, Burnham turned to Seattle-based freelancer **David Williams**. A trained geologist, Williams has worked as a ranger at Arches National Park in Utah and an instructor at Harvard University, where he taught programs on rocks and minerals. "I can't imagine not thinking about the landscape around us," he says. "Geology centers me."

Many trips to Cape Cod and the nearby islands over the years made New York City science writer **Beth Livermore** a good choice to explore the dunes, moraines, and kettle ponds of the cape. "I was struck by the contrasts between the cape's serene, present-day face and the dramatic story of its birth during the Ice Ages and its eventual death beneath the waves," she says.

Similarly, Denver-based writer and editor **John Murray** has had long experience with the beautiful cliffs and valleys of the Blue Ridge Mountains of Virginia and the cool, quiet chambers of Mammoth Cave in Kentucky. He is the author of more than 25 books on a wide range of natural-history topics and the editor of the Sierra Club's nature writing annuals.

For **Alianor True**, it was only a short detour from her Ann Arbor, Michigan, home to visit Niagara Falls. "I was immediately struck by the immensity and power of the falls, which contrast greatly with the landscape of the Great Lakes." **Mel White**, a natural-history writer from Little Rock, was another who didn't have far to travel when he explored Arkansas' Blanchard Springs Caverns. "The cave is one of the best in the country, and its setting in the Ozarks means the scenery above ground is just as beautiful as that below."

Nature and travel writer **Conger Beasley**, based in Colorado Springs, is the author of more than a dozen books on the West, including guides to the national parks of the Rocky Mountains. His long experience in the West led to his writing the chapter on the Badlands of South Dakota. And for **Jeremy Schmidt**, an award-winning travel writer and author of several National Geographic guidebooks, Yellowstone was just up the road from his home near Jackson, Wyoming.

"Far from being static," says Southwest writer **Nicky Leach**, "rocks are alive with their own stories." She tackles the epic landscapes of several parks in the region, including Canyonlands, Zion, and Death Valley. Also in the Southwest is **Rose Houk**, a frequent contributor to the Discovery Travel Adventures series, who claims that "few places in the world can take you into 'Deep Time' like the Grand Canyon. The history of Earth revealed in its lofty cliffs puts our brief lives in perspective."

Two writers based in the San Francisco area covered places in Northern California that are on the "must-see" list of every geology-minded explorer. Like the Grand Canyon, Yosemite is often crowded with visitors, but **Glen Martin**, an environmental writer for the *San Francisco Chronicle*, still finds the landscape inspiring, with granite walls reaching to the sky. And back in the Bay Area, a vivid memory of the 1989 Loma Prieta earthquake brought Californian **Blake Edgar** to a new appreciation of the San Andreas fault. "We tend to forget that the motion of California's crustal plates both creates and destroys," he says, since the same tremors that can level a city also raise mountains. "Those who appreciate California's coastal scenery choose to put up with its faults, too," he notes wryly. Edgar edited *Discovery Travel Adventures: Dinosaur Digs* and is co-author with paleontologist Donald Johanson of two books, *From Lucy to Language* and *Ancestors: In Search of Human Origins*.

Getting the story on Mount St. Helens was no picnic, according to **Tim McNulty**. "I camped in a downpour and my walks with geologists were blustery, to say the least. When I compared my notes with recordings made in the field, the overwhelming sound on the tape was wind." Idaho writer **Julie Fanselow** explored an even bigger catastrophe than the eruption of Mount St. Helens: Ice Age floods that scarred parts of Montana, Idaho, and eastern Washington, creating an area known as the Channeled Scablands.

In Gustavus, Alaska, **Greg Streveler** and his wife live a semi-subsistence lifestyle on a homestead of their own construction next to Glacier Bay. For him, "its complicated and beautifully exposed bedrock makes every inlet and headland a lesson in our area's geologic history." Not the chill of a glacier but the fire of Earth's interior drew volcano writer and photographer **Donna O'Meara** to write about fiery Kilauea near her home in Hawaii. "To see a volcano in action – lava jetting skyward, steam hissing from cracks in the ground, acrid sulfur stinging your nose – is to experience what primal Earth must have been like."

Thanks to the many park rangers and naturalists who reviewed the text. Thanks also to members of the Stone Creek Publications editorial team – Edward A. Jardim, Sallie Graziano, Michael Castagna, Judith Dunham, and Nicole Buchenholz.

Water erosion creates small wonders like this chasm (above) in the Grand Canyon as well as such monumental works as the canyon itself.

Balanced Rock (below), an eroded sandstone formation in Utah's Arches National Park, "teeters" in the warm glow of the setting sun.

Beach cobbles (opposite) on the Maine coast were delivered by glaciers ages ago and slowly rounded by waves.

Preceding pages: Sunrise reddens the cliffs of Dead Horse State Park in Utah, with Canyonlands National Park beyond.

Following pages: Minerals dissolved in hot groundwater produce the colorful palette of Yellowstone's Grand Prismatic Spring.

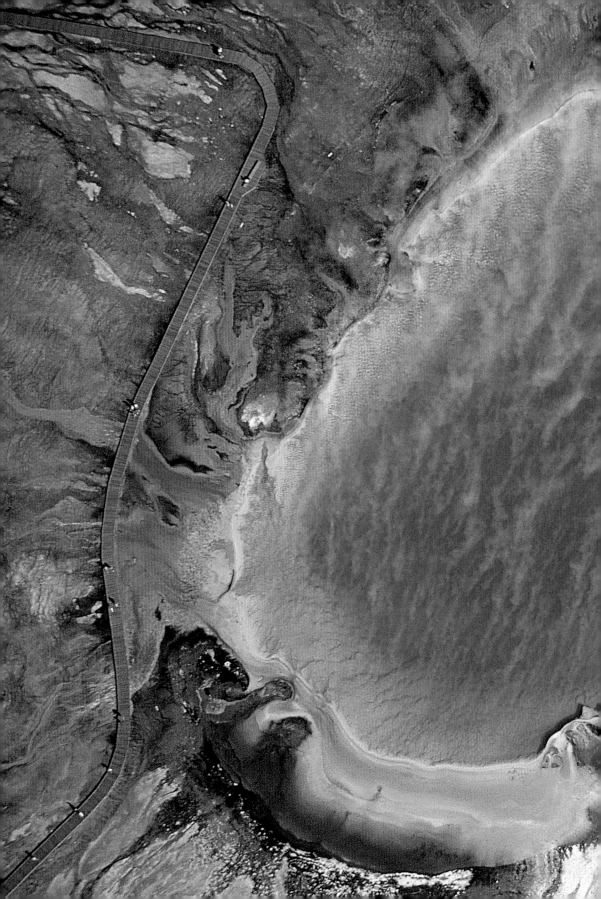

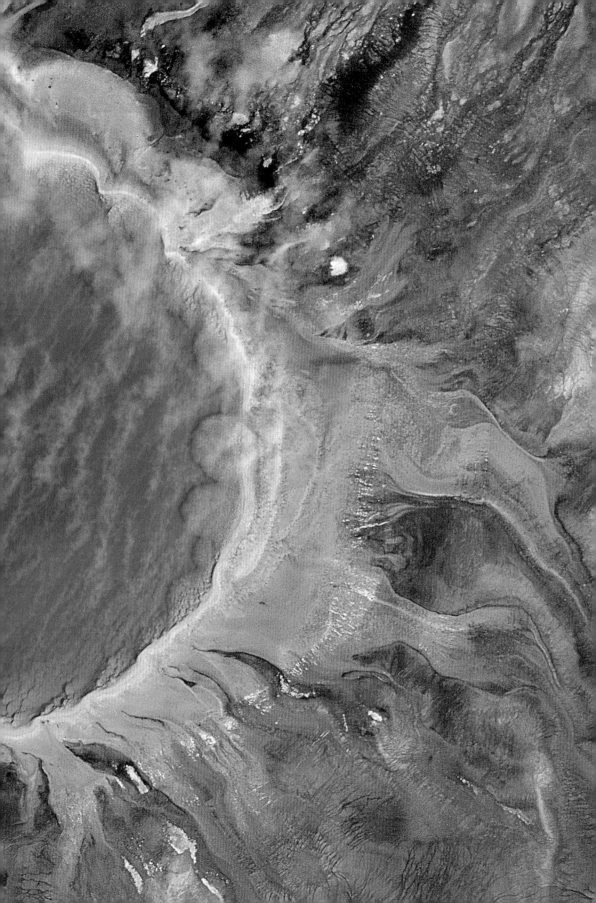

Table of Contents

MAPS

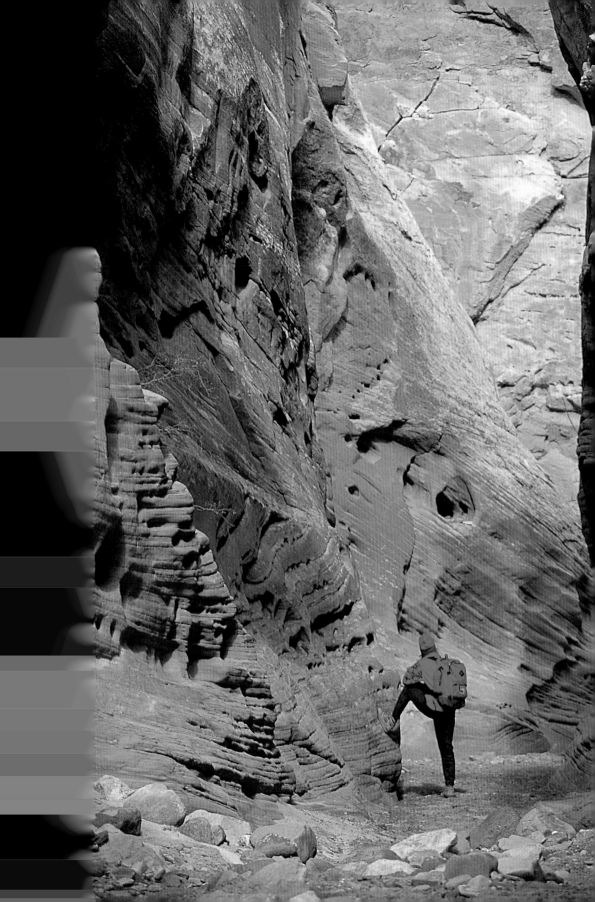

People with an interest in the natural world often focus on the here and now – a soaring turkey vulture, a bear fishing in a stream, a pod of migrating whales. The beauty and grace we find in living creatures touches a primordial nerve and fills us with delight. It's the immediacy we crave, the sense of connection with another living thing. ◆ Geology, on the other hand, requires a very different frame of mind. Grasping how the landscapes around us have taken shape demands that we understand large, impersonal processes which manifest themselves over great expanses of space and time. We must make a mental leap from the everyday bustle of humanity and embrace the rhythms of a planet that has only begun to register our presence on it. ◆ Many people are never able to make the transition. Geology appears on their radar screens only when it hits the headlines: an earthquake in California, a volcanic eruption in Washington state.

To understand the Earth on its own terms, one must first grasp the immensity of geologic time.

And yet, geology surrounds us – in the stone walls of our office buildings, the salt on our dinner tables, the hills and valleys we drive through every day. And not just us. The landscape influences weather and plays a major role in determining where plants and animals live. ◆ The first step for anyone with a serious interest in Earth science is to grasp the enormous depth of geologic time. The Earth, along with the rest of the solar system, is 4.56 billion years old. This is a huge number, virtually impossible to comprehend fully. But try comparing it to a more manageable time frame:

A slot canyon in Utah meanders through contorted beds of sedimentary rock.

Preceding pages: Whale's Belly, Paria–Vermilion Cliffs Wilderness, Arizona; Nevada Falls and Liberty Cap, Yosemite Valley, California; Lechuguilla Cave, Carlsbad Caverns, New Mexico.

Geysers (left) in Nevada's Black Rock Desert erupt when groundwater meets hot subterranean rock and bursts into steam.

Weathering attacks this rock on the California coast (below) in two ways: by physically chipping away its grains and by chemically dissolving its minerals.

Cavers (right) descend more than 500 feet into Ellison's Cave, Georgia.

If you were to collapse the age of the planet into a single year, then each day would equal about 12½ million years, an hour 521,000 years, a minute 8,700 years, and a second about 150 years. On this scale, if the Earth formed at midnight January 1, then the first durable crust appeared around February 17, microbial life appeared around March 3, and the atmosphere became rich in nitrogen and oxygen in mid-July. The first creatures to leave behind substantial skeletal remains appeared about November 5, the first animals crawled onto dry land around November 29, dinosaurs became extinct on December 25, and fully modern humans made it onto the scene at roughly 11:37 P.M. on New Year's Eve.

While geological events such as earthquakes and volcanic eruptions transpire relatively quickly, they are merely the culmination of processes that occur far more gradually. For example, the North Pacific tectonic plate is moving northwest along the San Andreas fault at about half an inch per year. At that rate you would need 44 million years to stroll from Los Angeles to San Francisco. Mountains can push up even more slowly, some rising 10,000 feet in 15 million years, or ¹⁄₁₂₅th of an inch per year.

Geology demands that we have faith in the power of time and in the grindingly slow accumulation of microscopic change into significant outcomes. Though it seems unlikely, water can wear down a mountain into a plain, continents can drift apart or collide, vast ice sheets can creep down from the poles, carbon can be turned into diamonds, trees can be transformed into stone.

Exploring the world of animals and plants reawakens us to the roots we share with all life forms. Earth science widens that perspective enormously, and reminds us that we all dwell within a vast natural story – that of the Sun, Earth, and solar system. This story began before life itself and will still be unfolding long after our species has vanished from the planet.

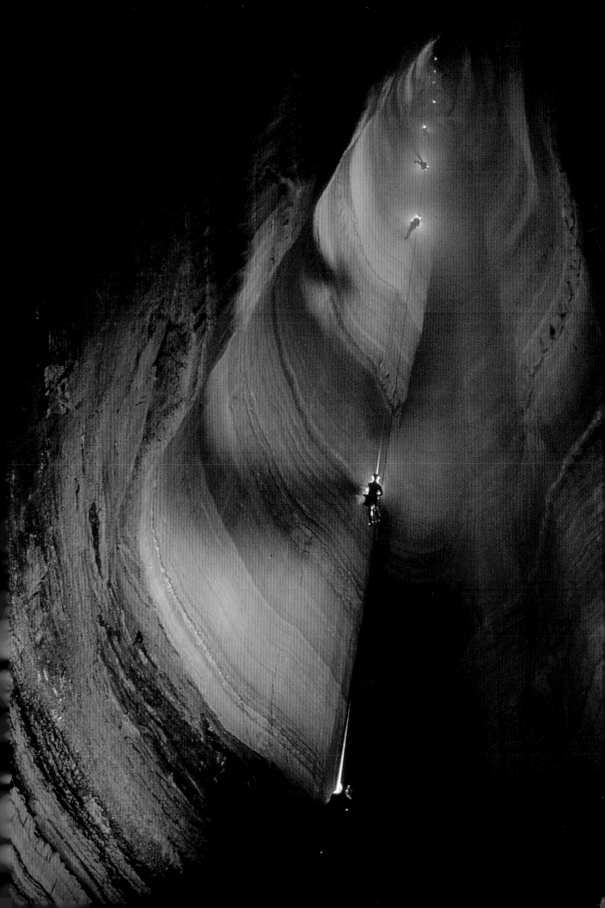

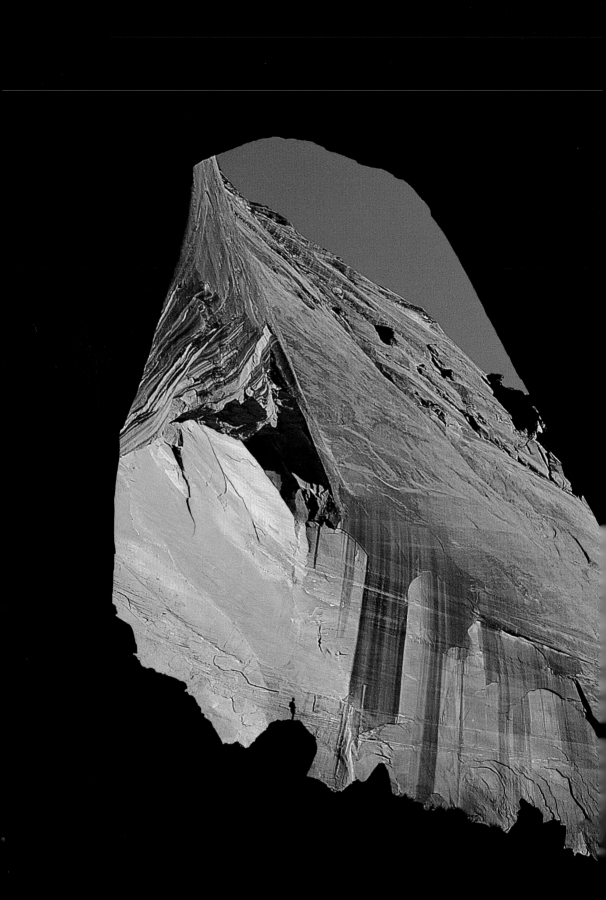

◆

Preparing For the Field

◆

Preparation is the key to a safe and enjoyable journey. Before hitting the road, familiarize yourself with basic geological concepts, invest in the necessary gear, and review the regulations governing rock collecting on public land.

Earth, like the solar system, was born 4.56 billion years ago, a date that can be fixed with some precision by the age of the oldest meteorites. During Earth's infancy, it bore scant resemblance to the terrestrial globe we know today. The third planet from the Sun came into being as an agglomeration of silicon compounds and iron and magnesium oxides, laced with smaller amounts of all the other elements. Neither continents nor seas existed. Life had yet to appear, and what atmosphere there was would not support life in any case. ◆ The embryonic planet took on added mass from the bombardment of countless planetesimals, small rocky bodies akin to asteroids orbiting in space. Ice-rich comets slammed into Earth as well, each delivering a wallop and a quantity of water. In time, Earth became the fifth largest of the Sun's nine planets, slightly greater than Venus, but only a quarter the size of Neptune. And it was a living body;

The face of the Earth is constantly changing as geologic forces simultaneously build it up and wear it away.

the impact of the planetesimals and comets, plus gravitational compression and the decay of radioactive elements, produced enough heat to create a global ocean of molten rock, or magma. ◆ Within 100 million years, the magma ocean had melted all the iron it encountered, allowing the metal to sink toward the center of the planet. As the iron descended, lighter elements rose, making layers within the sphere. At the heart of the planet was a dense core of solid and liquid iron. Surrounding that came a zone of partly molten heavy rock known as the mantle. And on top rested a thin crust of lighter rock. The crust itself consisted of two layers, a heavier one of basaltic rock

Volcanoes such as Washington's Mount St. Helens (left) let primordial heat leak from the Earth's interior.

Preceding pages: A natural arch frames a hiker in Arizona's Paria Wilderness.

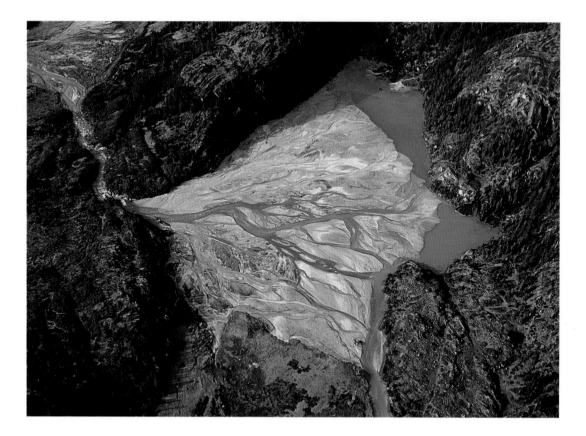

and a lightweight layer of granitic rock, which eventually shaped the continents.

Soon after Earth had separated into layers, it absorbed a cataclysmic, if glancing, blow from one of the remaining planetesimals orbiting the Sun. This wayward body – a piece of debris left over from the construction of the planets – was perhaps one-third the size of Earth. The collision blew off some of Earth's crust and upper mantle. Sailing into orbit around Earth, the remains of the planetesimal, together with ejected material from Earth, coalesced into a new body that began its own geological evolution. The Moon was born.

Meanwhile, gases escaped from the interior of the planet and started to form an atmosphere and hydrosphere. Important gases included hydrogen, carbon dioxide, and water vapor. Once the global sea of magma cooled, water vapor condensed to form the first oceans.

The Fires Below

Besides helping create the atmosphere and oceans, the separation of the planet into layers set up the driving force of geologic change: plate tectonics. This theory, put forward in the 1960s, divides Earth's crust – continents and ocean basins – into 10 large plates and several smaller ones, all of which are moving on the surface of the globe like self-propelled pieces of a monumental jigsaw puzzle.

CRUST

UPPER MANTLE

LOWER MANTLE

Convection Currents

OUTER CORE

INNER CORE

Fossils and the History of Life

A delta of "rock flour," or silt, left, is washed from under an Alaskan glacier.

Fossils like *Uintacrinus* (right) help reconstruct life in the oceans millions of years ago, while the fossils of mammoths, such as this deposit in South Dakota (bottom, right), tell about life during the most recent Ice Age.

Life on Earth began about 3.85 billion years ago. Since then, 95 percent of all organisms have gone extinct. The records of their passage are preserved in either the hard remains (bones, teeth, leaves, and seeds) or traces (burrows, tracks, and root impressions) they left behind.

Known as fossils, evidence of past life occurs primarily in sedimentary rocks and ranges in age from the earliest bacteria to mammoth bones less than 10,000 years old. There are three classifications: invertebrates, vertebrates, and plants. Invertebrates are by far the most common group. The first multicelled organisms appeared 700 million years ago and the first fossils with hard skeletal parts occurred 570 million years ago. These early animals, extinct for the last 245 million years, were all marine organisms. The earliest evidence of land-dwelling invertebrates comes from tracks left 440 million years ago by arthropods that resembled modern centipedes.

Vertebrates made their first appearance in the Ordovician period 500 million years ago. Less than an inch long, these first fish had no jaws and drew nourishment by filtering organisms from seawater. Modern-looking fish would not appear until 30 million years later, and four-legged amphibians would not crawl onto land for another 100 million years. Mammals go back 200 million years to primitive, mouse-sized creatures and evolved most of their forms by 25 million years ago.

The first plants to colonize land appeared 415 million years ago, yet so rapid was their evolution that all major forms had developed within 45 million years – all except flowering plants, which did not arrive for another 235 million years.

The energy that keeps everything in motion comes from decaying radioactive elements that produce heat within the Earth. As the heat rises, the mantle flows sluggishly in giant convection cells – basically, big bubbles of molten rock boiling slowly and eternally, like soup on the stove. The convection cells move the crustal plates, crunching them together and pulling them apart.

To see how Earth is shaped by plate tectonics, let's look at the North American plate and those that surround it. The North American plate includes all of North America plus Greenland and extends about a thousand miles into the Atlantic Ocean. Its neighbors on the east are the African and Eurasian plates; those on the west are the North Pacific plate along with the smaller Juan de Fuca and Cocos plates.

The contact zone between the North American plate and the African and Eurasian plates lies in the middle of the Atlantic Ocean, at the boundary of two of Earth's convection cells. These cells have been operating for roughly 200 million years. Before that time, these three plates were joined together (with others) into one massive supercontinent which geologists call Pangaea, meaning "all-Earth," and indeed, it covered two-fifths of the globe.

Pangaea itself was the result of earlier crustal collisions. A surviving memento of Pangaea's birth pangs are the Appalachian Mountains, which were thrust up 350 million years ago, when the edges of two continents merging into Pangaea buckled like the hoods of cars in a head-on collision. The impact produced mountains as tall as Everest, since reduced by

The swirling walls of a slot canyon (left) are eroded in bursts, with months of stillness broken by violent flash floods.

a plate made of less dense continental crust. When this happens, the denser material slips below the lighter rocks, diving into the mantle whence it came. Geologists call the process subduction. It has been happening for the past 36 million years off the west coast of the North American plate where the oceanic Juan de Fuca plate is colliding with continental crust in Washington state. As the Juan de Fuca plate descends, its leading edge melts in the high temperatures and pressures under the continent. The resulting magma rises and in places pierces the surface to form volcanoes such as Mount Rainier, Mount Hood, and Mount Shasta – not to mention Mount St. Helens, which erupted with awesome violence on May 18, 1980, ejecting nearly a cubic mile of rock and ash.

Another kind of plate-to-plate collision occurs a thousand miles south of the Juan de Fuca plate – at the storied San Andreas fault. In this contact zone the plates are sliding past each other, as the Pacific plate moves in a northwesterly direction against the North American plate. When the two plates cease to slide and lock up, the stress builds until the rocks reach their breaking point – and an earthquake results.

A map showing where earthquakes occur is in

erosion to less than one-quarter of their former glory.

Making an Ocean

Pangaea was not destined to last. The Atlantic Ocean

Subduction Zone

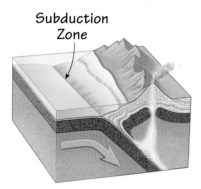

Transform Fault

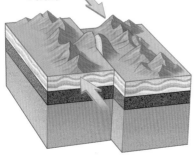

started to form about 200 million years ago, when a plume of magma reached the surface under part of the supercontinent. The force cracked Pangaea apart, splitting off the North American, African, and Eurasian plates. Driven by heat from the mantle, the magma spread between the plates and began to push them apart. They are still opening, or rifting, at a rate of about two inches per year, roughly twice as fast as your toenails grow.

As rifting continues, it creates a conveyor belt on the ocean floor. New basaltic ocean crust emerges at the rift and is carried away from its source. Eventually the oceanic crust collides with

Bump and grind. Where the Earth's crustal plates collide (left), one may dive under the other, returning to the mantle. Or rocks may seize up along a fault line, only to let go with an earthquake.

Farther downstream, the gradient lessens and the water flows onto the plains where the river meanders and develops a broad, flat valley. Finally, at the Gulf of Mexico, the great river spreads into a host of channels and deposits its load of sand and silt, building a delta into the ocean.

Sediment, when carried with the force of running water, acts like a belt of moving sandpaper, abrading and grinding the rock it passes over. Floods greatly enhance the ability of a river to carry sediments, and thus mold the landscape. Significant

essence a map describing the boundaries of Earth's crustal plates. And thousands of miles of plate boundaries translate to over a million tremors per year worldwide, the vast majority of which are, fortunately, registered only on seismographs.

At the source, water flows fast and clear, cutting a V-shaped valley, the characteristic shape of a river in steep descent.

Force of Water

Plate tectonics explain Earth's macrocosm, but there are other systems of geologic change constantly at work. These are the forces of erosion: water, wind, and ice. Water is observably the dominant sculptor on Earth. It carves canyons, creates caves and sinkholes, and erodes shorelines. Streams and rivers, tides and waves are its agents. Consider the Mississippi-Missouri River system, which rises in the Rocky Mountains.

Rocks rule the flow of water (top), but water always wins, as even the hardest rocks eventually erode, given enough time.

Underground erosion (right) produces caves and weirdly beautiful deposits of the mineral calcite, such as these in Carlsbad Caverns.

floods can move house-sized boulders and completely alter the topography of a valley in a relatively short time.

The layer of silt deposited by water will eventually compact into sandstone, and here is where you can read the geological history of the eons. The running water will leave ripple marks in the stone; other characteristic stream-produced features include graded bedding (larger grains topped by progressively smaller grains) and inter-mixed coarse and fine grains, which reflect floods washing sediments onto a floodplain.

Water running under-ground in limestone-rich areas has a powerful erosive capa-bility; acid in the groundwater dissolves the limestone, creat-ing sinkholes and caverns.

Ice and Wind

Ice is another major sculptor of the landscape in northern latitudes and at high eleva-tions. Two kinds of glaciers leave their mark: alpine and continental. Alpine or valley glaciers hone peaks into horns and ragged ridges known as aretes, while scooping out precipitous, basin-like cirques. The debris is pushed down-slope, creating piles of rubble called moraines. Typically, such glaciers carve out deep, U-shaped and dramatic hanging valleys. The Teton Mountain range in Wyoming and Glacier Bay in Alaska offer spectacular displays of alpine glacial features.

Continental glaciers have been in retreat since the end of the last Ice Age 12,000 years ago. At their maximum 20,000 years ago, vast sheets of ice up to two miles thick inched down from the north and ground across the North American continent, creating the gently rolling hills, small lakes, and fertile fields of Wisconsin, Minnesota, Iowa, and Michigan. As the glaciers advanced, boulders at the bottom of the ice sheet raked the bedrock like teeth. Kettle lakes were created by ice blocks calving off the retreat-ing glacier. Terminal moraines, like Cape Cod and Long Island, marked the end point of a glacier's forward progress and the beginning of its retreat. On the way back north, glaciers often deposited single large rocks called erratics that had been trans-ported far from their point of origin (Plymouth Rock is an erratic). And drumlins, com-posed of piles of rubble left behind by previous glaciation, were molded by the passing ice into elliptical hills running parallel to the direction of the glacier's movement; Beacon Hill in Boston is a well-known example.

Wind is the last of Earth's landscapers as it moves sediment from one place to another. In arid areas, wind created dunes from untold

Dawn streams through Keyhole Arch in Kansas (top), much as the water that created the arch once did.

A sandstone ribbon (right) spans 180 feet in Utah's Natural Bridges National Monument.

Alpine glaciers (left) break rock apart by brute force and by repeated freezing and thawing.

away in the same way that moving water will carve a canyon. Anyone who has been in the desert on a windy day knows that flying sand can sting. Such particles chip away easily at rock formations, breaking them down grain by grain, until a proud outcrop vanishes on the wind.

The pulse of the planet beats in all tempos. We can glimpse its largest-scale workings in places like Kilauea on Hawaii, where magma from the mantle bursts out onto the surface. And we can explore its smaller-scale action at sites like Mammoth Cave, where the ceaseless drip of water testifies to the process that shaped the underground wonder. The story of Earth began ages before we arrived on the scene and will continue long after we are gone. Yet even if we are just brief visitors, we can still explore its sights and delight in its marvels.

billions of bits of material, primarily quartz grains, although the dunes at White Sands in New Mexico are composed of gypsum. Dunes generally have a shallow-sloped, wind-blown face, which ends abruptly in a steep, leeward drop. As a

dune migrates, this leeward face creeps downwind, driven by the prevailing breezes; in light winds a dune may move only 10 feet annually, but where winds are strong and steady, it may migrate as much as a foot a day.

Wind can also cut rock

R eading the story of a rock is a little like reading a book. You begin with the basic plot line, which in the case of rocks means their mineral composition. Sometimes it's a simple tale of a single native element, such as gold or silver. More often, the narrative takes you on a fascinating excursion among the complexities of chemical compounds. ◆ Geologists have identified no fewer than 3,700 minerals, and the possible amalgamations are virtually infinite. However, only 30 or so minerals are common, and a scant 10 of these make up more than 90 percent of Earth's crust. In combination with each other or with other minerals, they compose most of the rocks around us. ◆ Geologists place rocks into one of three groups – igneous, sedimentary, or metamorphic – based on how they were formed. Igneous and metamorphic rocks are about equally common across the continents and make up roughly a third of all rocks. The floors of the ocean

Contained within every rock is the story of its creation.

basins are igneous, with a coating of sand, mud, and ooze that washed off the continents or settled out of the ocean water. The remaining two-thirds of continental rocks are sedimentary. ◆ Igneous rocks form by the cooling and solidification of molten rock, known as magma. If magma reaches the surface while hot, the resulting rock is termed extrusive or volcanic. Lava is a typical volcanic rock. However, if an igneous rock cools and solidifies within the Earth, geologists call it an intrusive or plutonic rock; granite is the most familiar example. ◆ How can you tell which is which? Look at the texture. Extrusive rocks like lava and basalt generally cool so fast that it takes a high-powered hand-lens to make out the individual minerals. But intrusive rocks such as granite cool slowly enough that they have time to

Schist, seen here in Utah's Westwater Canyon, is a metamorphic rock, gradually altered by immense heat and pressure deep within the Earth.

grow well-defined minerals that can be seen clearly with the naked eye.

Spotting granite in the field is relatively easy. It's usually gray, tan, or pinkish. Most of it has visible crystals of milky feldspar and translucent quartz with additional flecks of shiny mica and dull black hornblende. A freshly broken piece of granite can look startlingly like a scoop of vanilla ice cream shot through with micro-chips of chocolate.

Among the extrusive rocks, basalt makes up 70 percent of the rock on Earth's surface, most of it under the oceans. Basalt is generally black and filled with numerous cavities, or vesicles, the remains of trapped gas bubbles. It may have a ropy surface called

pahoehoe (PAH-hoy-hoy) or a sharp, blocky texture called aa (ah-ah). And it can flow like molasses across enormous swaths of terrain – as witness the 25,000 cubic miles of basalt that covered the Columbia Plateau of Washington and Oregon 20 million years ago. Other large basalt flows

include the Deccan Plateau in India, the Ethiopian Plateau in Africa, and the dark "seas" on the Moon.

Recycled Rocks

It is an axiom of geology that nothing lasts forever. Every natural feature will eventually succumb to weathering and

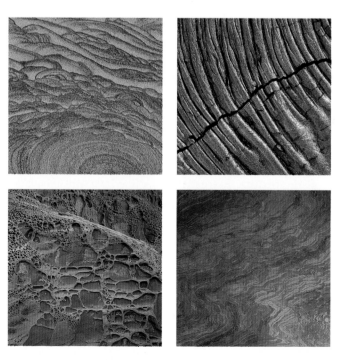

Texture and color vary greatly, especially if the rock is weathered. Clockwise from upper left: "wonderstone" sandstone, pahoehoe lava, iron-bearing jasper, and honeycombed sandstone.

Calcite columns (left) in Luray Caverns, Virginia, are made of calcium carbonate deposited, drop by drop, by seeping groundwater.

A scientist (right) displays a fossil fern. The rules governing the collection of rocks and fossils depend on where they are found.

be carried away bit by bit. Wind and water transport sediment downslope and deposit it at the bottom. Over tens of thousands to millions of years, this material accumulates and converts into sedimentary rock.

Geologists classify sedimentary rocks by grain size and composition. Sandstone, shale, and limestone are the most common. As the name suggests, sandstone is made up of sand-sized particles visible to the naked eye. By contrast, the sediments in shales are too small to see except with a hand-lens, and limestones consist of microscopic grains of calcium carbonate precipitated from seawater. Other sedimentary rocks include conglomerates, which are mixtures of sediments ranging from microscopic to boulder-sized, and evaporites left behind when briny water dries up.

Color in sedimentary rocks depends on their mineral composition and the amount and type of iron present. Oxidized iron (rust) creates shades of red, while iron from a low oxygen environment results in a greenish hue. Carbon turns rocks gray or black. White rocks lack iron and are composed mostly of quartz or feldspar.

The Rules of Collecting

While national parks prohibit the collecting of rocks, other types of public lands throughout the United States permit "rockhounding" and are often less developed and crowded. No matter where you are, however, it's critical to determine ownership. Private land is strictly off-limits unless you get permission beforehand. State parks and reserves may allow rock collecting, but rules vary from one state and location to another.

Rockhounds are welcome on public land administered by the U.S. Bureau of Land Management or U.S. Forest Service: These areas require no permits and don't limit amateur rock collecting as long as you don't plan to sell your finds or disturb the ground with explosives or power tools. Be careful not to collect on a marked mining claim; otherwise, you may be considered a claim jumper.

Fossil collecting has its own, more restrictive set of rules. For details, check with land management agencies in the areas where you plan to collect.

Sandstones are found in rivers, lakes, shallow marine estuaries, on beaches and in deserts, everywhere sand is deposited by wind or water. Because they are tightly cemented together, sandstones often form cliffs and ledges. A layer of sandstone may have been laid down by a single storm or over the course of thousands of years. Ancient sand dunes that have turned to rock display a characteristic called cross-bedding that reflects the cascading of wind-blown sand down the dunes' leeward faces. Examples of sandstones include the wind-deposited Wingate sandstone in Utah and the stream-borne Portland Formation in Connecticut, the rock that built the "brownstones" of Boston and New York.

Shales are also widespread. These fine-grained sediments, rich in clay, indicate quiet deposition in lakes or ocean bottoms. Grays and reds are typical shale colors, and the rock usually weathers into slopes and rounded hills. After a rainstorm, clays may absorb so much water that the shale turns into a sticky mess known as gumbo. Strikingly colorful shales occur in the Badlands of South Dakota and the Petrified Forest of Arizona.

Limestone is easy to distinguish with a simple test. If a drop of dilute hydrochloric acid makes the rock fizz, it's most likely limestone. Limestones are often rich in fossils, which tells you that the site was once under water. It may have been just a spring carrying calcium-rich water, as at Mammoth Hot Springs in Yellowstone

Field Photography

Taking good geological photos is no easy matter. A disposable camera is usually good enough for a simple record of something you've seen. But going beyond ordinary photos and getting something really striking calls for a little more technique.

Shooting around sunset and sunrise enhances low-contrast rocks. On a gray day, look for colorful elements like wildflowers or lichens that can brighten an outcrop. Sharp details are important, especially with minerals, fossils, or rock textures. Use a tripod and a high f/stop (f/11 or greater) to keep everything in focus. And give your photos a size reference by including a coin, a pocket knife, or a person in the scene.

Point-and-shoot cameras are adequate for many situations, but extreme close-ups call for an SLR – a single-lens reflex camera. SLRs can use macro-focusing lenses, and since the viewfinder will show you what goes on the film, you'll frame your shots more accurately.

Filters (if you have an SLR) are often helpful. A UV filter cuts through haze, and a polarizer reduces glare on water and other reflective surfaces. An 81A or 81B filter will enhance yellow, red, and orange colors while diminishing bluish shades. A graduated neutral-density filter, dark on top and lighter toward the bottom, compensates for bright backgrounds and darker foregrounds – for example, shooting a lava flow against clouds.

Film? The pros choose slides because they are less subject to processing variations, but prints are often more practical for amateur geologists. Avoid "store brand" films; Kodak and Fuji have excellent films in both print and slide formats. Slow speeds (ISO 100) are great for sunlit scenes, but many photogs opt for ISO 400 or faster, trading a little graininess to gain versatility.

Caves pose special challenges, yet offer chances to create stunning photos. High film speeds – at least ISO 400 – are essential, and slide film works better than print film. Humidity is usually a problem; keep a waterproof bag handy to protect the camera from mud and dampness. Wrapping your hand around the lens just before you shoot may help prevent fogged optics. A small tripod, if allowed in the cave, lets you "paint the rocks" using a flashlight during a time-exposed shot. Otherwise, use the camera's flash but be prepared for somewhat stark and contrasty results.

Capturing impressive images seems fairly easy in the West's dramatic landscapes (above), but striking photos can be made anywhere.

National Park – or it may have been a tropical ocean where masses of marine organisms accumulated, as in the 220-million-year-old reef in the Guadalupe Mountains of Texas.

Limestone is also a cave-forming rock. Water percolating through limestone can quickly erode openings and passageways. Good examples include Mammoth Cave in Kentucky and Carlsbad Caverns in New Mexico.

Metamorphic Rocks

When rocks of any kind are subjected to great pressures and temperatures, their minerals often recrystallize or realign, essentially creating a new rock. Shale, for example, turns into slate, basalt into schist, and granite into gneiss. Rocks can be metamorphosed on a massive scale, as during a continental collision, or on a much smaller scale by the intrusion of magma, which bakes the surrounding rock.

Slate often tells a two-part story. For example, the Monson slate of Maine is a common roofing material in the eastern United States. This black rock originated as a fine-grained sediment in a deep ocean basin and eventually solidified into shale. The basin was then crushed by tectonic forces which compressed the shale into slate. This forced the randomly oriented clay minerals, which are flat and sheet-like, into a rigidly parallel arrangement – something like gathering a collapsed house of cards into a tight deck.

Marble is metamorphosed limestone or dolomite, and

is favored by sculptors and architects. Easily worked, it is a gleaming white in its purest form, but minor imperfections can create every shade imaginable. Two other metamorphic rocks, schist and gneiss, are often found in the cores of once deeply folded mountain belts. Among the most ancient of rocks, they commonly go back hundreds of millions to billions of years. Schist's characteristic texture of visible, aligned mica sheets make it easy to distinguish in the field. Equally easy to spot, gneiss shows distinct banding of light and dark minerals; it is frequently used as a building stone.

Environments and Time

Looking at rocks in the big picture, geologists speak of formations. The term describes either a single rock layer or a series of similar layers found over a broad geographic area. Formation names have two parts: the first refers to where the layer was found, and the second part indicates the type of rock – Chuckanut sandstone, for instance, or Salem limestone. Geologists also use the word formation for a suite of related layers, such as the Morrison Formation, a sequence of Jurassic sandstones and shales 180 million years old that occurs from Utah to Nebraska.

Wherever two different

formations come into contact, it indicates a shift in the environment. If two rock layers touch but were formed at widely different times, geologists speak of an unconformity. One such unconformity in the Grand Canyon represents 1.4 billion years of missing information.

You can't help wondering what happened in that yawning interval. But the record is erased – gone for good. Indeed, wherever you look, there's always more rock missing than visible. Erosion never rests. It is constantly

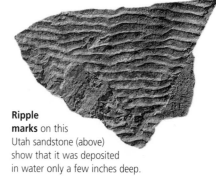

Ripple marks on this Utah sandstone (above) show that it was deposited in water only a few inches deep.

breaking down and carrying away rock. All we can do is sort through the remains and try to rebuild a geological jigsaw puzzle that is missing more than half its pieces.

Igneous rocks usually lack internal layering and often weather into the rounded shapes seen in this New Hampshire streambed (right).

Identifying Rocks

IGNEOUS

Granite

Granite is formed of magma that has cooled slowly underground. It varies in color, appearing white, gray, pink, or red. Crystals, visible to the naked eye, are medium- to coarse-grained and consist of minerals such as feldspar, quartz, biotite, and mica.

The rock resists weathering in humid climates, but tends to crumble easily in desert conditions. Massive granite bodies slough off sheets of rock, a process known as exfoliation.

Peridotite

Like granite, peridotite is made of magma that has cooled gradually underground but, unlike granite, is composed mainly of olivine and pyroxene. It is greenish to dark gray or nearly black in color. Crystals are medium- to coarse-grained and visible to the naked eye. The rock is often found in dikes (a seam of magma injected into place under high pressure) and sills (a massive, horizontal sheet of magma). Exposed surfaces weather to a pitted or crumbly texture with rounded contours. Dunite, a form of peridotite that is nearly pure olivine, has a light green to yellow-green color.

SEDIMENTARY

Conglomerate

Conglomerate rock forms when deposits of sand and gravel in stream beds or shorelines are buried, and rock fragments are cemented together by silica or calcite. Color varies widely, depending on the rocks that comprise the specimen. Particles vary in size from microscopic to a quarter-inch or larger. Conglomerate often crumbles when struck with a hammer. It is sometimes called pudding-stone; when the rock fragments are angular, it is termed breccia.

Sandstone

Sandstone forms when beaches, sand dunes, or shallow-water deposits of sand are buried and cemented together by silica or calcite. It is composed mainly of small grains of quartz, which are visible to the naked eye. Colors include white, gray, tan, yellow, pink, and red. When struck with a hammer, sandstone breaks around the quartz grains. Weathering may reveal cross-bedding, ripple marks, or other features of the original environment.

METAMORPHIC

Gneiss

Like all metamorphic rocks, gneiss is formed when existing rocks recrystallize under great heat and pressure. Composition varies widely depending on the parent rock; gneiss can form from granite, shale, sandstone, or previously metamorphosed rock. Colors resemble those of granite – white, gray, pink, red. Crystals form parallel streaks or bands; they are medium- to coarse-grained and visible to the naked eye. Outcrops of gneiss weather much like granite, but display the characteristic banding and streaking.

Schist

The composition of schist varies depending on the source rock, which may include shale, sandstone, rhyolite, basalt, or previously metamorphosed rock. It is silvery white, gray, greenish, or brown in color. Crystals are medium- to coarse-grained and are visible to the naked eye. Schist has a shiny or soapy luster, and tends to split into sheets when struck with a hammer.

Basalt

Basalt is made of magma that erupted on the surface and quickly cooled. It consists mainly of feldspar, pyroxene, and olivine, and its crystals are usually too small to be seen even with a magnifying glass. The rock is black in color but may weather to reddish brown in humid areas where iron, found in pyroxene minerals, oxidizes or rusts. If the magma is rich in gas, basalt may be riddled with bubbles called vesicles. Flows of basalt with a ropy texture are known as pahoehoe lava; flows that are rich in silica have a rough, angular texture and are called aa lava. Basalt is also found in dikes and sills, and may form columnar joints as it cools and shrinks.

Rhyolite

Rhyolite is created when granitic magma erupts at the surface instead of cooling slowly underground. It is composed mainly of feldspar, quartz, biotite, and mica, and its crystals are visible to the naked eye or with a magnifying lens. Colors range from light gray, tan, and yellow to reddish. Rhyolite is found in lava flows, and may display flat or swirling bands that resemble sedimentary layers. Pumice is extremely porous rhyolite.

Shale

Shale, also known as mudstone, consists of clay grains that have been deposited on lake- or seabeds; it may grade into a fine sandstone (if small quartz grains are abundant) or limestone (if calcium carbonate grains are common). Colors are gray, brown, or black; individual particles are too small to see without a magnifying glass. Shale is usually deposited in thin layers and may display mud cracks, which indicate that it formed in very shallow water.

Limestone

Limestone is made of calcium carbonate produced by mollusks, corals, and other forms of marine life or precipitated out of seawater and deposited on the seafloor. Particles are usually too small to see without a magnifying glass but may include visible shells or spherical grains. Colors range from white to gray to tan. Limestone weathers readily in humid areas but is highly durable in arid conditions. The rock fizzes when it comes in contact with hydrochloric acid or strong vinegar.

Quartzite

The various colors of quartzite - white, gray, tan, yellow, pink, and red - resemble those of sandstone, its parent rock. Its crystals are fine-grained and not as distinct as those of sandstone. It breaks into sharp edges when struck with a hammer and, unlike sandstone, it splits across the grains rather than around them. Quartzite is a hard rock and resists erosion. It usually forms in massive deposits and weathers very slowly.

Marble

Marble is mostly white but may shade into gray, tan, red, green, or even black. The crystals are too small to see with the naked eye, but in rare cases some fossils may be visible in the rock. Composition is generally calcium carbonate, although some forms may contain dolomite (magnesium carbonate). Marble bubbles when it comes in contact with hydrochloric acid but not as vigorously as limestone, its parent rock.

Outfitting yourself to explore the geology of planet Earth has a lot in common with any kind of outdoor adventuring. If you're an experienced hiker or nature-walker, you know the importance of well-fitting boots and proper clothing. Basically, you want to dress in clothes that are both comfortable and rugged enough for scrambling around rocks. ◆ There are a few specialized items that you probably don't have already. These include a hand-lens, rock hammer, notebook, and compass. A hand-lens ($10 to $50) with a magnification of 10x is crucial if you want to identify minerals. You use a lens by holding it close to your eye and bringing up the rock sample until it's in focus. Be sure to attach a brightly colored shoelace or string to keep the lens from going astray; better yet, wear the hand-lens around your neck, or tie it to a belt loop. ◆ A rock hammer ($25 to $40) is almost a membership badge for geologists (although you can't use one in state and federal parks, where collecting is generally forbidden). The most popular style is solid steel and has a sharpened point for picking and prying, plus a blunt end for pounding and breaking rock. Get one with a nylon- or rubber-covered handle to help absorb some of the impact. Estwing makes fine hammers and leather holders you can attach to your belt. If you get more involved, you may want a rock chisel and small sledge hammer for extracting samples from outcrops. A sledge with a two- or three-pound head is perfectly adequate. And don't forget plastic goggles to protect your eyes from flying rock chips. ◆ To tag and carry your collected samples, bring along some newspaper and a roll of masking tape. Attach a small piece of tape to the sample to note where you

The proper equipment will help ensure a safe and rewarding experience in the field.

Some geological exploring calls for special training and techniques, but in most cases the equipment is simple and relatively inexpensive.

The Brunton Pocket Transit (above) can be used to determine "strike," "dip," and the direction of north.

collected it; more tape will keep it wrapped snugly in the newspaper.

A notebook is essential. It can be simple or fancy, and what you record in it is up to you. Many geologists prefer those with waterproof pages. You can choose either lined or unlined paper: Lined paper helps organize your notes, but it also gets in the way of sketches. A medium-sized notebook (about 5 x 8 inches) fits in a coat pocket. Many geologists, both professional and amateur, also use a camera to supplement their notes and sketches.

Don't Get Lost

The routes described in this book are usually well-marked, but for much backcountry "geologizing," a compass is a necessity. A fluid-filled compass with a calibrated outer ring can be used to measure the directional trend of a rock. This is technically known as the "strike." The "dip" of a rock bed is the angle of tilt, or the plunge, that a fold makes into the ground. Strike and dip help you envision what the rocks are doing in three dimensions, all part of getting the picture.

The standard geologist's compass is made by the Brunton Company and combines a compass and tilt meter for measuring strike and dip. It's known as a Brunton Pocket Transit and costs about $200. Ordinary compasses run from $20 to $50. A recent addition for some explorers is a global positioning system (GPS) receiver, which helps in plotting the location of rocks on topographic maps. Along with a compass, a GPS unit also helps if you

Road-Cut Geology

Road cuts are to geologists what X-rays are to doctors – a way to peer under the skin and see the structures inside. Each time a road slices into a hill, it opens a window into the area's geology. Road cuts range from small rises along a shoulder to towering cliffs that display a whole range of otherwise invisible folds, faults, and layers.

For information about road cuts and other geologic features in a region, check out the guidebooks published by local geological associations. These usually have maps for road trips highlighting nearby formations. State geological surveys are another good source of information. Cvancara's *A Field Manual for the Amateur Geologist* provides phone numbers and addresses for such surveys. Also check the guides in the *Roadside Geology* series published by Mountain Press.

When you stop to explore a road cut, use extreme care. Road cuts always have loose rocks, so watch out for what's above you, especially if it's rained recently. Traffic is another hazard. Park as far off the road as possible and in a place where oncoming drivers can see the car (and you). Always park on the side where you plan to explore – and never, ever try to cross a highway on foot!

Collecting rocks from road cuts is subject to state or federal law, and is usually forbidden without a permit. As you explore a road cut, you may see markings painted on the rocks. No, not "Jason loves Katie," but cryptic notations like "97-Tr-135u." These indicate places where geologists have taken samples for their research. Please don't disturb.

Big road cuts can be spectacular. This one on I-70 near Denver, Colorado, below, chops through the Morrison Formation.

get carried away looking at rocks and end up lost.

A few final items round out your field gear. Dilute hydrochloric acid carried in a small polyethylene bottle helps to determine if a rock is limestone. The acid reacts with the calcium carbonate to produce carbon dioxide, which fizzes. (Strong vinegar also works but is less reactive.) A knife with a locking blade is useful for prying out fine material and for determining mineral hardness.

Otherwise, you'll want to take along the usual outdoor gear: water bottles, sunscreen, a wide-brimmed hat, lightweight binoculars, and a day-pack. (Keep the pack smallish or you'll be tempted

to bring home too much.) Pack a basic first-aid kit with plenty of Band-Aids; many rocks have sharp corners and edges, and it's easy to pick up scratches. Watch for loose rocks in road-cuts and in the field; they can tumble without warning. Also keep an eye on your fellow explorers, who might inadvertently dislodge material above you.

As your senses become more attuned to the geology around you, one last danger awaits: erratic driving. Keep at least one eye on the road. When you feel a pang of regret as you watch a particularly splendid outcrop dwindle in the rear-view mirror, you'll know you've become a geotourist.

A rock hammer with a blade-shaped pick (above) often works better with sedimentary rocks than a pick-pointed hammer.

A bike courier's bag (below) holds as much as a day pack and is usually easier to carry and set down.

◆

Destinations

◆

Whether you explore soaring mountains, deep canyons, bizarre subterranean passageways, or rugged badlands, experiencing America's geological wonders will forever change your perspective on the landscapes around you.

Niagara Falls
New York-Ontario

CHAPTER 4

N iagara Falls isn't the highest or widest waterfall on Earth. But it is certainly the best known and arguably the most impressive. The first sight of Niagara, like the first glimpse of the Grand Canyon, usually amazes even those well-accustomed to natural wonders. When Theodore Roosevelt first visited Niagara Falls in 1879 at the age of 21, even the normally loquacious future president – a perpetual chatterbox, from all reports – was moved to silence. What is more, while many of the world's great waterfalls lie in wilderness, Niagara Falls is located at the cusp of two North American cities (one in Canada, the other in the United States), both named for the falls. Together they are host to more than 10 million visitors each year. ◆ Many American travelers make their base in **Buffalo, New York**, about 20 miles southeast of the falls. From the American side, the best views of the 2,100-foot-wide **Horseshoe Falls** (which carries 90 percent of the flow) and the 1,075-foot-wide **American Falls** can be had at **Niagara Reservation State Park**, just off the Robert Moses Parkway. Visitors can walk to numerous

Few places better illustrate the erosive power of water than these mighty cascades on the Niagara River.

overlooks along this most spectacular segment of the 33-mile **Niagara River**, which runs from Lake Erie to Lake Ontario. On the Canadian side, **Niagara Falls, Ontario**, is larger than its namesake across the gorge and offers all levels of lodging and tourist services. **Queen Victoria Park** lines the rim of the gorge, and at Table Rock provides excellent – even slightly frightening – close views of Horseshoe Falls. ◆ The story of the falls begins in the Ice Age. As the vast sheets of continental ice melted away about 13,000 years ago, water gathered inland in low areas to form immense glacial lakes. In time

Viewing Niagara Falls from below is an exciting, if somewhat soggy, experience that gives you a feel for the river's power.

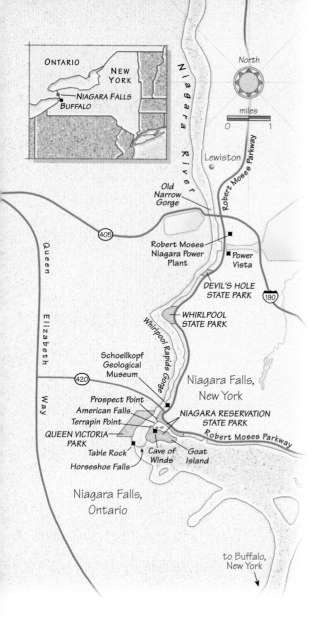

when warm subtropical seas covered North America. The uppermost layer is composed of hard, dense Lockport dolomite. Beneath that comes a dark layer of Rochester shale, then several thin limestone layers followed by a band of Grimsby sandstone. Farther down lies Power Glen shale, Whirlpool sandstone, and finally Queenston shale at the bottom. Everything erodes more easily than the uppermost dolomite, so the normal state of the falls is a lip of dolomite projecting over a recessed wall.

The erosion accelerates as Niagara River water percolates into the Lockport dolomite along joints and cracks in the rock and undermines the softer sediments below. From time to time, pieces of dolomite crumble away into the gorge, and the falls retreat upstream toward Lake Erie. Because of its tremendous flow, Horseshoe Falls is being eaten away at the rate of a foot or two each year, or about 2,000 feet since the falls separated two millennia ago. By contrast, American Falls, with its weaker flow, has lost only a few dozen yards. Another function of flow is the 160-foot-deep plunge pool that the falling water has scoured out at the base of Horseshoe Falls. American Falls has no such pool; in fact, the flow is not sufficient to break up and carry away the huge pile of talus that reduces the height of the falls to about half that of the Canadian side.

Niagara Gorge below the falls has its own set of strikingly different faces. In the Whirlpool Rapids Gorge and the Old Narrow Gorge near the power plants, the passage is narrow and the river shallow, which squeezes the water into a hydrologic frenzy. Elsewhere, the gorge broadens, allowing the water to assume a more placid flow.

Human requirements also play a critical role in the state of the falls. Since the river is important to Canada and the United States for both electric power and tourism, a 1950 treaty established parameters for the flow of water. During "tourist time" – basically, 8 A.M. until late at night, from April 1 through October 31 – the falls discharge 100,000 cubic feet per second, about half their natural flow. At other times and continuously from

these became the four upper Great Lakes: Superior, Michigan, Huron, and Erie, which now collect the drainage from an area larger than Sweden. Almost all of that water – more than 200,000 cubic feet per second – goes over Niagara Falls or through its electric power generators.

Unlike most big falls, Niagara does not drop in steps but plummets unhindered in one 167-foot dive. The reason is that the most durable of the Niagara rock layers is on top. The falls expose flat-lying strata of sedimentary rock laid down around the Ordovician-Silurian boundary 440 million years ago,

The **Aerocar** (right) takes visitors on the Canadian side down to the Whirlpool.

Horsehoe Falls, in the distance (below), carries most of the river's flow, while just 10 percent goes over American Falls, left of center.

November to March, this flow is reduced further, to 50,000 cubic feet per second, as more water is sent into the generators.

Caves and Whirlpools

Most people begin their tour on the American side with a stop at the **Schoellkopf Geological Museum**. Built where a power station was destroyed by a rockfall in June 1956, the museum is a couple of hundred yards north of the river and can be reached via Robert Moses State Parkway. Museum displays feature the geological history of the Niagara River; the rocks, minerals, and fossils of the Great Lakes region; and the development of hydroelectric power from 1920 on. Near the museum is **Prospect Point**, which has an observation tower with a splendid vista of the American and Horseshoe Falls. After visiting the museum and Prospect Point, many people take a ride on the *Maid of the Mist*, which motors surprisingly close to the base of the falls for a spectacular view from below.

Other popular places to visit include **Goat Island**, **Terrapin Point**, and **Cave of the Winds**. Goat Island, the small island that divides the river into the two falls, is reached

Horseshoe Falls (left) has eroded a hole in the river bottom as deep as the falls are high.

The Cave of the Winds walkway (right) takes visitors near the foot of the Bridal Veil section of American Falls.

American Falls (below), illuminated by floodlights, is an eerily beautiful sight.

by footbridge from Prospect Point. On the island, Terrapin Point sits near the edge of Horseshoe Falls, while Cave of the Winds lies at the base of the Bridal Veil Falls section of the American Falls.

Three miles downstream is **Whirlpool State Park**, with its view of rapids that truly have to be seen to be believed. In their ferocity, the churning whirlpool and surrounding white water evoke the dreaded whirlpools encountered by Homer's Odysseus. The Whirlpool, believed to be 126 feet deep, has a curious feature. Normally, it circles counter-clockwise, but when the water is low it reverses direction.

Two final areas bear mention – **Devil's Hole State Park**, just south of the Power Vista, and the **Power Vista** itself. The former is an impressively deep basin cut into the upper Lockport dolomite. It carries historical significance as the spot where a hundred British troops were ambushed and massacred in 1763 by a party of Seneca Indians fighting for the French.

The Power Vista is part of the **Robert Moses Niagara Power Plant**. Here you will find a wonderful view of the Niagara

Gorge – and something special inside the visitor center: a marvelous mural by Thomas Hart Benton depicting Jesuit Father Louis Hennepin who, in 1678, was the first European to view the falls.

TRAVEL TIPS

DETAILS

When to Go

Summer in the Niagara Falls area is warm and pleasant, with highs in the 80s, though nights are frequently cool. Winter temperatures dip into the 10s and 20s, with heavy snowfall. Fall offers cool but comfortable weather, colorful foliage, and fewer visitors.

How to Get There

Commercial airlines serve Buffalo, New York, which is about 30 minutes away by car, and Toronto, Ontario, about 90 minutes away. Car rentals are available at the airports. Amtrak, 800-872-7245 or 716-285-4224, stops in Niagara Falls, New York; Via Rail stops in Niagara Falls, Ontario.

Getting Around

An automobile, convenient for reaching Niagara and exploring the region, is an encumbrance when sightseeing downtown. A shuttle, the Niagara Parks People Mover, offers all-day service on the Canadian side. Niagara Transit's city buses serve the U.S. side. When crossing into Canada, U.S. citizens should carry photo identification and a birth certificate or voter registration card. Some non-U.S. citizens will need to present a passport and/or visa. Canadians entering the United States should have a passport and a birth certificate and photo identification. For details, call Canada Immigration, 905-354-6043, or U.S. Immigration, 716-885-3367.

INFORMATION

New York Tourism

1 Commerce Plaza, Albany, NY 12245; tel: 800-225-5697 or 518-474-4116.

Niagara Falls, New York, Convention and Visitors Bureau

310 4th Street, Niagara Falls, NY 14303; tel: 800-421-5223 or 716-285-2400.

Niagara Falls Tourism, Ontario

5515 Stanley Avenue, Niagara Falls, ON L2G 3X4, Canada; tel: 800-563-2557 or 905-356-6061.

CAMPING

Four Mile Creek State Park (Lake Road, Youngstown, NY 14174; tel: 716-745-3802 or 716-745-7273) has hiking trails and more than 200 tent and trailer sites with electricity on the shore of Lake Ontario. It's open late April to mid-October. Numerous commercial sites are available on both the U.S. and Canadian sides; contact the visitor bureaus for a list.

LODGING

PRICE GUIDE – double occupancy

$ = up to $49 $$ = $50–$99

$$$ = $100–$149 $$$$ = $150+

Bedham Hall

4835 River Road, Niagara Falls, ON L2E 3G4, Canada; tel: 905-374-8515.

The 1875 house is situated on the Niagara River Parkway within walking distance of the falls. The inn has four air-conditioned guest rooms with antiques, fireplaces, and queen-sized or double beds; three of the rooms have private baths. Gourmet breakfasts are served; lunch and dinner are available upon request. $$–$$$

Manchester House Bed-and-Breakfast

653 Main Street, Niagara Falls, NY 14301; tel: 800-489-3009 or 716-285-5717.

Less than a mile from the falls, and three blocks from the geological museum, this bed-and-breakfast has three guest rooms with private baths. Twin and queen-sized beds are available. The house was built in 1903 and renovated prior to the inn's opening in 1991. A sitting room has an organ, piano, television, and gas fireplace. $$

Michael's Inn by the Falls

5599 River Road, Niagara Falls, ON L2E 3H3, Canada; tel: 800-263-9390 or 905-354-2727.

Situated one block north of the Rainbow Bridge, this four-story motel overlooks the Niagara Gorge and River. Many of the motel's 130 rooms have a view of the falls. The inn also offers "Jacuzzi theme rooms." $$–$$$$

Park Place Bed-and-Breakfast

740 Park Place, Niagara Falls, NY 14301; tel: 800-510-4626 or 716-282-4626.

Within walking distance of the falls, this late-19th-century house was the home of James G. Marshall, the founder of Union Carbide. The inn's four guest rooms have period antiques and private baths. The Arts and Crafts Movement house has hardwood floors, wood-paneled walls, and Oriental rugs. Gourmet breakfasts are served. The living and dining rooms have fireplaces. $$–$$$

Red Coach Inn

2 Buffalo Avenue, Niagara Falls, NY 14303; tel: 800-282-1459 or 716-282-1459.

Built in 1923, this Tudor-style house overlooks the Upper Rapids on the Niagara River, near the north end of American Falls. The inn's 11 guest rooms have private baths and antique furnishings. Most of the rooms have fireplaces and kitchenettes. $$–$$$$

TOURS

Cave of the Winds

Niagara Reservation State Park, Goat Island, Niagara Falls, NY 14304; tel: 716-278-1770.

An elevator ride descends to river level, where wooden walkways guide visitors to the foot of American Falls.

Journey Behind the Falls

Queen Victoria Park, Niagara River Parkway, Box 150, Niagara Falls, ON L2E 6T2, Canada; tel: 905-354-1551.

Elevators descend to tunnels that lead to terraces beside and behind Horseshoe Falls. Rain gear is provided.

Maid of the Mist

151 Buffalo Avenue, Niagara Falls, NY 14303; tel: 716-284-8897 or 905-358-5781.

Boats depart every 15 minutes from two docks, one at Prospect Point on the U.S. side, the other at Clifton Street and Niagara River Road on the Canadian side. Tours on the Niagara River culminate at the foot of the falls. Rain gear is provided.

Niagara Helicopters

3731 Victoria Avenue, Niagara Falls, ON L2E 6V5, Canada; tel: 800-281-8034 or 905-357-5672.

Helicopter tours of Horseshoe and American Falls and the Whirlpool last about nine minutes.

Rainbow Air

454 Main Street, Niagara Falls, NY 14301; tel: 716-284-2800.

This service offers helicopter tours of the falls. Hours are seasonal; call for schedule.

MUSEUMS

Niagara Falls IMAX Theatre

6170 Buchanan Avenue, Niagara Falls, ON L2G 7T8, Canada; tel: 905-358-3611.

The theater's six-story-high screen presents the story of Niagara Falls, including daredevil barrel riders and tightrope walkers.

Schoellkopf Geological Museum

P.O. Box 1132 (Robert Moses Parkway), Niagara Falls, NY 14303.

Built at the ruins of a power station destroyed by a 1956 rockfall, the museum concentrates on the geology of the Niagara River.

Excursions

Adirondack Mountains

Adirondack Park Visitor Center, Route 30, P.O. Box 3000, Paul Smiths, NY 12970; tel: 518-327-3000.

These wild, scenic mountains sprawl across 10,000 square miles of upstate New York. Geologically, the Adirondacks are part of the Canadian Shield, a vast expanse of ancient rock that underlies much of the continent. The range is a large dome of Precambrian metamorphic rock surrounded by lower Paleozoic strata, uplifted some 100 million years ago and now much eroded. The highest peak, at 5,344 feet, is Mount Marcy. Exhibits at the visitor center cover the region's geology and ecology.

Niagara Escarpment

Bruce Peninsula National Park, P.O. Box 189, Tobermory, ON N0H 2R0, Canada; tel: 519-596-2233

East of Niagara Falls, the rock layer that created the famed landmark soon disappears under younger sediments. But to the northwest, a pronounced escarpment of Niagara limestone and dolomite winds for 450 miles across southern Ontario, vanishing into the blue waters of Lake Huron and Georgian Bay at the tip of the Bruce Peninsula. The escarpment can be seen between Hamilton and Guelph on Highway 6; from Owen Sound, Wiarton, and Dyer's Bay; and on hiking trails in Bruce Peninsula National Park, near Tobermory, Ontario.

Sudbury Basin

Science North, 100 Ramsey Lake Road, Sudbury, ON P3E 5S9, Canada; tel: 705-522-3701

Almost two billion years ago, a very large meteorite struck the Earth and left a crater perhaps 120 miles across. The impact was strong enough to force material up from the Earth's crust. As a result, the Sudbury area is host to one of the largest mining complexes in the world. Today, the remains of the crater, much deformed by later tectonic activity, lie at the edge of the Canadian Shield, an old, heavily eroded expanse of crystalline rocks. Displays at Science North in Sudbury feature exhibits on the meteor.

Cape Cod and the Islands
Massachusetts

CHAPTER 5

For multitudes of summer visitors, Cape Cod is a good place to loll on the beach and work on a tan. But for those with a sense of landscape, this arm-and-fist peninsula extending 35 miles into the Atlantic Ocean has a fascinating geological tale to tell. The cape and its associated islands were laid down by Ice Age glaciers – and immediately came under such assault by wind and water that they will one day disappear into the sea. The signs are everywhere.   ◆   The first acts of this planetary drama were played out in a blink of time's eye. The great Laurentide Ice Sheet started forming in Canada 75,000 years ago, when the climate was cold enough to prevent winter snows from melting during the summer. As the layers thickened, eventually reaching 10,000 feet, the snow compacted into ice and the entire mass began to slide across North America and onto the continental shelf. About 20,000 years ago, the Ice Sheet arrived at its maximum extension, poking three immense fingers, or lobes, out into the North Atlantic.   ◆

Bulldozed into existence by an ice cap that covered much of North America, this sandy peninsula is slowly vanishing under the attack of the ocean.

There it sat for several thousand years, until the Earth began to warm again and the Laurentide glacier commenced its retreat. In the first stage, the ice melted back to where the islands of Nantucket and Martha's Vineyard now lie. In the second stage, the ice withdrew to where the cape and its westward-reaching string of Elizabeth Islands reside. Each melt-back left behind vast heaps of clay, sand, gravel, and boulders. This rubble forms the moraines, or hummocky backbone, of Cape Cod and the Islands. From the forward edge of the glacier, meltwater streams carried gravel and sand outward to produce the outwash plains, or flatlands.

The cliffs at Gay Head on Martha's Vineyard are Cretaceous and Tertiary clays bulldozed into place by glaciers some 20,000 years ago.

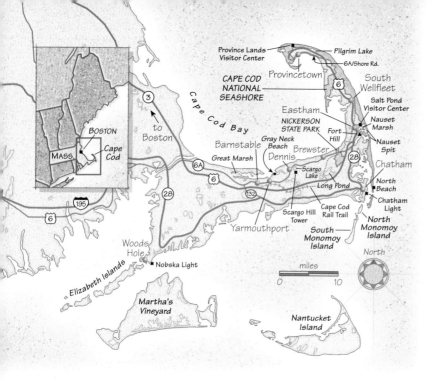

back then, you might have seen mastodons foraging on the tundra. "Fishermen still bring me mastodon teeth that they've dredged up in their nets," says Oldale, who has been studying the area for 30 years.

Geologist Oldale suggests that the best place to start your journey back through the ages is at the southwestern tip of the cape. This is the oldest part of the glacially-created landscape, prominent with moraines, kettle ponds, and outwash plains. Then, as you head east and around the elbow to the northern tip, you'll encounter newer formations such as sea cliffs, barrier beaches, salt marshes, and sand dunes.

Rolling Moraines and Ice-Block Lakes

At **Nobska Light** on Shore Road in Woods Hole, you'll be able to see the large rolling moraines of Martha's Vineyard just a few

But it wasn't until about 6,000 years ago, when sea level rose, that the south shore looked like it does today. Until then, says Robert Oldale, of the U.S. Geological Survey, "worldwide sea level was approximately 300 feet below its present level." And instead of Nantucket Sound, a broad coastal plain stretched south to the edge of the continental shelf. Standing on the cape

miles away across Vineyard Sound. To your right, the moraines of the Elizabeth Islands stretch away into the distance.

Now drive north to **Scargo Hill Tower** on State Route 6A in **Dennis**. A short circular staircase leads to the top of the tower, where you'll enjoy a sweeping view of the entire north shore: Provincetown to the right, the mainland to the left, and **Cape Cod Bay** straight ahead.

About 19,000 years ago, before the sea broke through and filled Cape Cod Bay, the entire expanse was fresh water. What geologists call Glacial Lake Cape Cod pooled here, fed by meltwater from the retreating glaciers. The immense lake, extending from about Scituate just south of Boston, 30 miles east to Provincetown, stood 60 to 80 feet above present sea level. Where it drained, it cut river valleys – today's Cape Cod Canal was originally a natural spillway from this lake to the south shore.

Sitting directly below the tower, **Scargo Lake** is a kettle pond, or ice-block lake. When the ice lobes retreated, chunks of ice, some more than a mile across, cracked off. Sediment washing off the glacier buried many of these. When the blocks melted, they left hollows a hundred or more feet deep, many reaching below present sea level. At least

500 such freshwater ponds dot the cape. A ride on the **Cape Cod Rail Trail** through **Nickerson State Park** in **Brewster** will give you a close-up view of **Long Pond**, which at 743 acres is one of the area's largest.

Building Up, Tearing Down

You'll start seeing newer formations that also characterize the cape at **Chatham** on the southern point of the elbow. Climb to the viewing platform of **Chatham Light** on Shore Road for a glimpse of the **Cape Cod National Seashore**, a 30-mile expanse of creamy sand. Farther out stand two great barrier beaches, **North Beach** and **Monomoy Island**. Together, they tell a dynamic story of beach building and erosion on the cape.

No sooner had the ice sheet retreated north than the sea began to sculpt the glacial deposits. First, sand and gravel were eroded and redeposited to build beaches. Then sand drifting along the shore created sandbars and barrier beaches wherever the coastline took a sharp turn. North Beach in fact once spanned the entire eastern horizon before you. But nothing on the cape is permanent.

In January 1987, North Beach was breached by a violent storm. "Since then," says Oldale, "the opening has widened to more than a mile and is migrating southward, exposing more and more of the upland to wave attack." To appreciate the power of the ocean, descend the staircase at the parking lot to the beach below and walk north at

Atop Gay Head's clays (left) lies a veneer of glacial sand and gravel. Slumps show where blocks of cliff are slipping into the sea.

Erratic boulders (above, right) were carried by glacial ice from other parts of New England. Many show scratches from the glacier's movement.

low tide. Here you will see septic tanks, wells, and bits of house foundations where summer homes stood before 1987. If you have time, consider a boat trip to Monomoy Island. Monomoy was once a single sand spit, but in February 1978 a winter storm split it in half. Seals now bob like corks in the water that flows between the two island halves.

Washed to the Sea

A visit to **Fort Hill** in **Eastham** is worthwhile for its network of sanctuary walking trails. And from there you'll have a sensational view, framed by wave-cut cliffs, of Nauset Spit and Nauset Marsh, both of which are geologically interesting.

Nauset Spit, like North Beach, has undergone dramatic changes over the years. Between 1605 and 1978, the southern end was displaced half a mile toward land. The northern end once had sand dunes as high as 40 feet before a storm in 1978 wiped them totally flat. Nevertheless, for all the storms and stresses, the spit remains to shelter a healthy, long-lived salt marsh.

Indeed, **Nauset Marsh** is flourishing in the quiet waters behind the spit, where a tide-driven flow of sediment keeps the bottom shallow. At high tide, wide channels separate large islands of swaying cordgrass. At low tide, the muddy banks reveal all sorts of creatures, from periwinkle snails and fiddler crabs to minnows and grasshoppers. These attract birds – gulls, sandpipers, ducks, seaside sparrows, northern harriers, and short-eared owls. Moreover, all this is readily accessible to visitors. There are two walking trails – one from the parking lot and one from **Salt Pond Visitor Center**. Paddlers can rent sea kayaks and canoes,

Agassiz the Icebreaker

The **White Cedar Swamp** trail (opposite, top) in Wellfleet leads from dunes to swamp forest.

Nauset Marsh (opposite, bottom) formed behind a barrier spit of land that protects it from ocean waves.

Louis Agassiz (below) was the first to recognize that glaciers had covered much of Europe and North America in geologically recent time.

When Jean Louis Rodolphe Agassiz was a boy, the Swiss Alps where he lived filled his days with all manner of enjoyment. After he became a scientist, they got to be serious business. The Alps inspired him to formulate his Ice Age theory: that glaciers once covered parts of the Earth and helped to sculpt it into the world we know today.

Louis Agassiz, as he was known, was born in 1807. He was more than a geologist; he was interested in all living things and their environments. Early in his career, he studied with Georges Cuvier, taught zoology at the University of Neuchatel, and was noted for his work on fossil fish. Later, he wrote the multivolume *Contributions to the Natural History of the United States* and founded the Museum of Comparative Zoology at Harvard College, where he was a professor.

Beginning in 1837, Agassiz focused his formidable intellect on glaciers. That summer, in front of the Swiss Society of Natural History, he argued that there had been a prehistoric Ice Age, during which glaciers extended from the North Pole to the borders of the Mediterranean and Caspian Seas. The evidence, he said, was in the rubble they left behind. His ideas were greeted with disbelief, so he set out to prove them.

Agassiz established a camp on Unteraar glacier in Switzerland to study the glacier's motion and in 1840 published one of the first treatises on glacial movements and deposits. He subsequently searched for signs of glacial erosion and sedimentation all over North and South America and Europe – research that established beyond question that the Pleistocene epoch of two million to 11,000 years ago was punctuated by periods of intense glaciation, with warmer interglacial intervals between them.

and join tours run by local outfitters. Other marshes to explore are the **Great Marsh** in **Barnstable**, which spans 4,000 acres along Route 6A, and **Gray's Neck Beach** in **Yarmouthport**, with a quarter-mile boardwalk.

The Newest Cape

The **Province Lands Visitor Center**, off Route 6 near **Provincetown**, is the place to marvel at giant sand dunes. The visitor center has an indoor/outdoor observation deck offering a 360-degree view of nothing but silky-looking undulations, dune grass, and ocean. This is the youngest, most changeable part of the cape. In fact, once you pass **Pilgrim Lake** on your approach, you have left the glacial cape behind. The whole area was built by sand washed from the southern cape and blown by the wind.

Most Cape Cod sand dunes reach heights of no more than 20 or 30 feet. "But," says Oldale, "on Provincetown Spit some dunes, like Mount Ararat, are more than 100 feet high." And they are forever changing. While soil and vegetation stabilize some dunes, most are moving inland, thanks to the wind. They are filling Pilgrim Lake and blow across Route 6 like snowdrifts. Several times a year, plows must push the sand back to keep the road open. Meanwhile, the waves keep eating away. A glance at the map shows that the outer cape is narrowest near South Wellfleet. Here's where the ocean may someday break through. This would turn the northern cape into an island. Still, "that won't happen for a long time," says Oldale. "Eventually, the cape and the islands will lose their battle with the sea, but the sea will not win easily. It'll take thousands of years before the cape is reduced to shoals and low-lying islands, and perhaps 5,000 or 6,000 years to be completely drowned."

TRAVEL TIPS

DETAILS

When to Go

Early fall and late spring are the best seasons to explore Cape Cod; the weather is pleasant and the towns and beaches are much less crowded than in summer. Temperatures in October range from 47° to 59°F; in May, from 48° to 62°F. Summer highs are usually in the mid-80s; nights are often cool and damp. Winter temperatures are in the 30s and 40s.

How to Get There

Many airlines serve Logan International Airport in Boston (two hours from Hyannis); commuter airlines link Logan to Provincetown Airport and Barnstable Municipal Airport, both on the cape. Bonanza Bus Lines, 800-556-3815, runs between Boston and Hyannis. Amtrak, 800-872-7245, provides limited rail service to Hyannis. Ferries serve Cape Cod and nearby islands year-round.

Getting Around

The cape is best traveled by automobile. Rental cars are available at the airports, except Provincetown.

Handicapped Access

Two visitor centers in Cape Cod National Seashore, Salt Pond and Province Lands, are wheelchair-accessible, as are many overlooks and most bathrooms in the park.

INFORMATION

Cape Cod National Seashore

99 Marconi Site Road, Wellfleet, MA 02667; tel: 508-349-3785. Province Lands Visitor Center, Race Point Road, Provincetown, MA 02657; tel: 508-487-1256

Cape Cod Chamber of Commerce

Routes 6 and 132, P.O. Box 790, Hyannis, MA 02601; tel: 888-332-2732 or 508-862-0700.

CAMPING

Nickerson State Park, 508-896-3491, has 420 campsites in Brewster. For a list of privately operated campgrounds in Cape Cod National Seashore, contact the Cape Cod Chamber of Commerce.

LODGING

PRICE GUIDE – double occupancy

$ = up to $49	$$ = $50–$99
$$$ = $100–$149	$$$$ = $150+

Beechwood Inn

2839 Main Street, Barnstable, MA 02630; tel: 800-609-6618 or 508-362-6618.

Built in 1853, this authentically restored Queen Anne-style house features wood-paneled walls, a pressed-tin ceiling, and period antiques. The bed-and-breakfast has six guest rooms with private baths and refrigerators; two rooms have fireplaces, some offer views of Cape Cod Bay. A wraparound veranda is furnished with rockers, gliders, and wicker furniture. The inn, within walking distance of historic Barnstable Village, serves gourmet breakfasts. $$–$$$$

Best Western Chateau Motor Inn

105 Bradford Street West, Provincetown, MA 02657; tel: 508-487-1286.

Decks at this hilltop motel overlook the waters surrounding Provincetown. Within two miles of the national seashore, the motor inn has 54 spacious rooms, each with a refrigerator and television. Gardens and a heated pool are on the premises. The motel is open May through October. $$–$$$$

Blue Dolphin Inn

Route 6, North Eastham, MA 02651; tel: 508-255-1159.

The Blue Dolphin, three miles north of Salt Pond Visitor Center, offers 49 rooms, all with patios. The inn is open April through October. $$–$$$

Captain's House Inn

369-377 Old Harbor Road, Chatham, MA 02633; tel: 800-315-0728 or 508-945-0127.

This bed-and-breakfast occupies an 1839 Greek Revival house, originally owned by a sea captain. Set on two acres, the inn has 19 guest rooms with antique furnishings and private baths. Most rooms have fireplaces and queen- or king-sized four-poster beds; some have whirlpool tubs and refrigerators. Picnic lunches are prepared upon request. $$$–$$$$

Chatham Highlander Motel

946 Main Street (Route 28), P.O. Box 326, Chatham, MA 02633; tel: 508-945-9038.

Set on a knoll amid a grove of pines, this modern motel has 28 ample rooms with ceramic-tile baths, televisions, and refrigerators; most rooms have two double beds. Open from May through October, the motel is a convenient stopping place for those eager to explore the cape's "elbow." Gardens and two heated swimming pools are on the premises. $$–$$$

Mariner Motel

555 Main Street, Falmouth, MA 02540; tel: 508-548-1331.

Affordable accommodations are available at this 30-room motel, conveniently located for those who want to explore the Woods Hole area or base their operations in the southwest corner of Cape Cod. All rooms have refrigerators. The motel is open March through mid-November. $–$$$

TOURS

Cape Cod Museum of Natural History

869 Route 6A, P.O. Box 1710, Brewster, MA 02631; tel: 800-479-3867 or 508-896-3867.

The museum leads hikes on North Monomoy Island, accessed only by boat. Overnight accommodations are available in the restored quarters of a lighthouse keeper.

Willie Air Tours

Provincetown Municipal Airport, Race Point Road, Provincetown, MA 02657; tel: 508-487-0240.

Passengers aboard a 1931 Stinson Detroiter see from the air how the ocean is shaping the outer cape's beaches and dunes.

MUSEUMS

Cape Cod Museum of Natural History

869 Route 6A, P.O. Box 1710, Brewster, MA 02631; tel: 800-479-3867 or 508-896-3867.

The museum has two floors of interactive natural-history displays, including the new Shape of the Cape. Three nature trails wind through 85 acres, which encompass wetlands, beech forest, and a salt marsh.

Woods Hole Oceanographic Institute Exhibit Center and Science Aquarium

15 School Street, Woods Hole, MA 02543; tel: 508-289-2663 or 508-289-2252.

This small center offers displays and videos about the institute's various research projects. Among the exhibited vessels and instruments is a full-sized model of a small research submarine. The Science Aquarium, at 166 Water Street, was established in the late 1800s and remains an enjoyable place to learn about marine life.

Excursions

Acadia National Park

P.O. Box 177, Bar Harbor, ME 04609; tel: 207-288-3338.

Devonian-age granite and diorite form the spine of Acadia National Park, set almost entirely on Mount Desert Island. Take the 27-mile drive on Park Loop Road for exceptional views of shoreline, forest, and mountains. Somes Sound, which runs north to south through the island, is the only true fjord on the New England coast. It was gouged out by a lobe of a Wisconsin-age glacier, which widened and deepened the natural joints in the rock.

Bay of Fundy

Fundy Geological Museum, 162 Two Islands Road, Parrsboro, NS B0M 1S0, Canada; tel: 902-254-3814.

The bay's famous tides leave eroded cliffs and sea stacks in their wake. Travelers interested in geology should not miss the north shore of Minas Basin and the Fundy Geological Museum in Parrsboro. Nearby cliffs reveal evidence of 325-million-year-old fossils. The museum displays some of the world's oldest dinosaur bones, models of ancient landscapes, rich mineral deposits culled from the area, and ancient tree and animal fossils from Coal Age forests.

Palisades Interstate Park

Alpine Approach Road, P.O. Box 155, Alpine, NJ 07620; tel: 201-768-1360.

Across the Hudson River from Manhattan is an impressive cliff of Triassic-age diabase, an intrusive volcanic rock. As the diabase cooled, it formed fluted, columnar joints whose appearance inspired the nickname Palisades. Hiking trails in the park, which extends about 12 miles north from the George Washington Bridge to the New York border, meander along the striking rock layer.

Blue Ridge Mountains
Virginia

To grasp the geological essence of the Blue Ridge Mountains, push firmly on a tablecloth or try surfing across the floor on a throw rug. The folds and buckles of the cloth illustrate what happened to the rocks of this Appalachian range hundreds of millions of years ago.　◆　The Blue Ridge Mountains, which form the northwestern edge of Virginia and are a segment of the larger Appalachian chain, are an ancient range that time has worn down to its figurative bones. The peaks and high gaps are made largely of crystalline Precambrian and early Paleozoic rocks once buried in the core of the mountains. What brought them to the surface was a combination of large-scale tectonic forces and simple erosion.　◆　The main tectonic movement occurred when the Atlantic Ocean opened more than 200 million years ago. A thick stack of sedimentary and igneous rock **The Appalachians expose** was shoved northwestward with such **their crooked spine along** force that some portions rode up and over others **one of America's most** along a thrust fault. These heaped-up layers **spectacular mountain roads.** then eroded almost completely. Washed toward the sea long ago, they now lie on the Piedmont, or coastal plain, and offshore on the continental shelf. What remains is mostly the massive, erosion-resistant basement rock. West of the Blue Ridge, the folded and buckled sedimentary layers roll toward the horizon. The ridges are mainly sandstone, with less resistant limestone and dolomite in the valleys.　◆　Perhaps the best way to view the region's geology is to follow **Skyline Drive**, a serpentine mountain road that runs about 105 miles along the spine of the Blue Ridge through **Shenandoah National Park**. Mileposts starting at the northern entrance simplify navigation, and dozens of overlooks, wayside exhibits, and hiking trails highlight significant features.

A pair of hikers surveys the Blue Ridge.
Tectonic forces and erosion have
exposed the core of the mountains
along some summits and ridges.

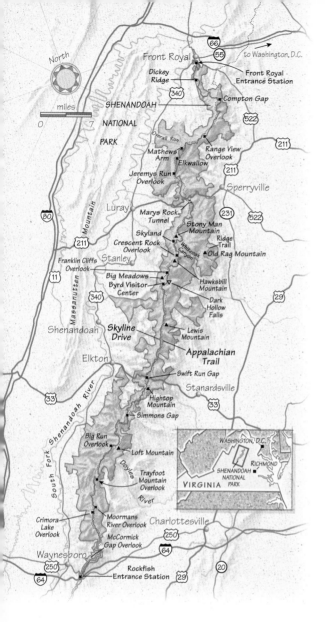

On the map:

North

miles
0   7

66
55 to Washington, D.C.
Front Royal
Dickey Ridge
340
Front Royal Entrance Station
Compton Gap
SHENANDOAH
522
NATIONAL
PARK
Overall Run
Mathews Arm
Elkwallow
Range View Overlook
211
211
Jeremys Run Overlook
Sperryville
Luray
Marys Rock Tunnel
231
522
Stony Man Mountain
Skyland
Crescent Rock Overlook
Ridge Trail
Whiteoak Canyon
Old Rag Mountain
Franklin Cliffs Overlook
Stanley
Massanutten Mountain
Big Meadows
Byrd Visitor Center
Hawksbill Mountain
29
340
Dark Hollow Falls
Shenandoah
Skyline Drive
Lewis Mountain
Elkton
Appalachian Trail
80
211
11
Swift Run Gap
Stanardsville
33
Hightop Mountain
33
Simmons Gap
Big Run Overlook
Loft Mountain
WASHINGTON, D.C.
RICHMOND
Doyles River
Trayfoot Mountain Overlook
SHENANDOAH NATIONAL PARK
VIRGINIA
South Fork Shenandoah River
North Fork Shenandoah River
Crimora Lake Overlook
Moormans River Overlook
Charlottesville
McCormick Gap Overlook
250
64
Waynesboro
250
64
Rockfish Entrance Station
29
20

Greenstone and Granite

The highest ridges and peaks of Shenandoah National Park are capped with greenstone or granite, both of which are highly resistant to erosion. The greenstone along the Blue Ridge – the Catoctin greenstone – is easily recognized by its pale gray-green to yellowish-green color. The degree of greenness depends on the amount of various minerals present in the rock. Greenstone is a dense, crystalline, metamorphic rock that forms when basalt (or lava) is subjected to high temperatures and pressures within the Earth. Catoctin greenstone sometimes contains narrow veins of light-colored quartz and calcite injected into the rock during the metamorphic process.

Hawksbill Mountain (near mile 47.5), 4,051 feet in elevation, is topped with exposed greenstone. Elsewhere, huge weathered boulders of granite dominate mountain summits, as on 3,268-foot **Old Rag Mountain** (near mile 48.0).

Hikers on **Ridge Trail** on Old Rag Mountain pass a particularly interesting feature, where a stairlike formation in a dike of greenstone has weathered faster than the flanking walls of granite. The granite found in the Blue Ridge Mountains is generally sandy white to dark gray, although in places it appears pink from iron-bearing minerals. Radiometric dating has determined that the Old Rag granite, which makes up the enduring core of the Blue Ridge Mountains, formed deep in the Earth 1.1 billion years ago.

There's more to Blue Ridge geology than greenstone and granite, however. To the north and south of **Jeremys Run Overlook** (mile 26.4) are sedimentary rocks such as pebble conglomerates, slate, and sandstone that form the cap of the Blue Ridge in this area. Looking west from many overlooks in the northern reaches of the park, you'll see the long profile of **Massanutten Mountain**. Its crest is sandstone, and it has large talus fields strewn with boulders below. These are the result of frost-wedging (the expansion of ice in cracks, prying open the rock) and heaving that occurred during the ice ages, when the climate here was like the Arctic today (although glaciers never crept this far south).

Hikers often see evidence of ancient volcanic events in the Shenandoah highlands dating back some 800 million years, well before the tectonic collision that raised up the Appalachians. One of the most easily observed is the profile of the "old man's face" on 4,011-foot **Stony Man Mountain** (near mile 41.7); the "face" eroded from an outcropping of lava. Another volcanic feature is **Big Meadows** (near mile 51), a 119-acre treeless area that is part of a five-square-mile plateau at an elevation of about 3,550 feet.

Big Meadows rests upon a massive 1,800-foot Precambrian lava flow called the Catoctin Formation, which is not visible on the surface. This formation also includes the metamorphic greenstone mentioned above and sedimentary layers of fragmented sandstone and volcanic debris. Geologists believe that the big lava deposit was formed by at least a dozen separate flows. As the lava cooled, it contracted into a vast network of polygonal cracks and columnar joints. The stockadelike columns in the lava can also be seen at **Compton Gap** (near mile 10.4), **Franklin Cliffs** (just north of Big Meadows), and **Crescent Rock** (near mile 44.4).

When **Marys Rock Tunnel** was blasted out by highway engineers in the 1930s, a striking Catoctin feeder dike was

exposed. It formed when lava surged upward through a narrow fissure in the granite about 600 million years ago. The six-foot-wide dike appears as a dark band toward the right of the tunnel's northern entrance.

Waterworks

Among the best-known features of the Blue Ridge Mountains are the numerous waterfalls. Particularly spectacular in spring, they gush with snowmelt and seasonal rain. The tallest, **Overall Run** (near mile 22.2), is almost 95 feet high. More than a dozen other waterfalls in the park drop more than 40 feet. These include the tremendous step-series waterfalls in **Whiteoak Canyon** (near mile 42.6) and the two great falls on the **Doyles River** (near mile 81.1). Most of them cascade over exposed benches of Catoctin greenstone. One of the best examples is **Dark**

Columnar jointing (opposite, top) is commonly found in outcrops of basaltic rock throughout the Blue Ridge area.

Dark Hollow Falls (right) tumbles across Catoctin greenstone, a metamorphosed basalt, near Big Meadows.

Hollow Falls, reached by a 1.4-mile round-trip trail just south of Big Meadows.

epic task long ago, but powerful natural forces continue to alter the features of the park. In 1995, for example, a rare "500-year" thunderstorm produced flash floods that scoured valleys, exposed new formations, eroded rocks, and reshaped waterfalls.

Wind, rain, and ice continue to sculpt and subtly refine the mountain scenery. A crack in a ledge gives water a chance to attack the rock. A lichen slowly erodes the surface of a stone with its secretions. A hiker taking an illegal shortcut in between switchbacks starts soil erosion. Even the acidified rain of the region has a geological effect, as it causes the less-resistant rocks to dissolve more quickly. The work of nature in altering the geology of the Blue Ridge Mountains will be complete only when the peaks have become a level plain – at which point a whole new cycle of change will begin.

It may seem as though the geological forces shaping the Blue Ridge finished their

Against the Grain

The first serious student of water gaps in northwestern Virginia was the nation's third president, Thomas Jefferson. In October 1783, he visited the confluence of the **Shenandoah** and **Potomac Rivers** near Harpers Ferry. There he viewed the deep pass, or water gap, where the river current had, over time, powerfully cut across the Blue Ridge Mountains.

Jefferson wrote in his book *Notes on the State of Virginia* (1787) that "the passage of the Potomac through the Blue Ridge is perhaps one of the most stupendous scenes in nature … [as] the mountain [is] cloven asunder." The other great water gap in the vicinity of Shenandoah National Park is that of the **James River** just to the south, near Jefferson's famous Natural Bridge, which he purchased and proudly owned all his life, even building a cabin near the unusual geological site.

These Blue Ridge water gaps represent an impressive display of the ability of moving water to carve through fairly resistant rocks. It is probably not an accident that the Potomac River formed a water gap north of the park, where the layers of Catoctin lava are less than 50 feet thick. Here the river did its erosive work faster than other streams that didn't have the geological advantage of softer rock. Farther to the south, where the subterranean lava beds are more substantial, it probably would have been impossible for the river current to maintain its cutting action through the slowly uplifting rock.

Marys Rock Tunnel (above) was blasted out of a granite mountainside in the 1930s.

Veins of granite and other igneous rock run through an exposed cliff face (right).

The Potomac River (below) flows through a water gap in the Blue Ridge at Harpers Ferry.

DETAILS

When to Go

Fall and summer attract the most visitors to Shenandoah National Park. Rain showers are common in summer; snow and ice, prevalent in winter, occasionally close Skyline Drive. Foggy, cool, and wet conditions may occur at any time. Temperatures in the park are approximately 10 degrees cooler than in the lowlands. High temperatures in summer average about 75°F; in winter, about 40°F. Nighttime temperatures are usually 20 degrees lower than daytime.

How to Get There

The park lies about two hours west of Washington, D.C. Commercial airlines serve Washington's Reagan National and Dulles Airports; commuter airlines serve Charlottesville-Albemarle Airport in Virginia. Interstate 66 leads west to Front Royal at the northern end of the park, U.S. Highway 211 crosses the middle of the park at Thornton Gap, and Interstate 64 passes the park's south end at Waynesboro.

Getting Around

An automobile is essential for touring Skyline Drive through Shenandoah; no shuttle, bus, or taxi service operates in the park. Car rentals are available at the airports.

Handicapped Access

Dickey Ridge and Byrd Visitor Centers are fully accessible, as are most overlooks. Limberlost Trail and portions of Big Meadows Trail are accessible. Picnic areas and campgrounds have accessible sites; inquire at park headquarters. Most bathroom facilities and buildings are accessible; some assistance may be required.

INFORMATION

Front Royal Chamber of Commerce

414 East Main Street, Front Royal, VA 22630; tel: 800-338-2576 or 540-635-3185.

Luray-Page Chamber of Commerce

46 East Main Street, Luray, VA 22835; tel: 888-743-3915 or 540-743-3915.

Shenandoah National Park

3655 U.S. Highway 211 East, Luray, VA 22835; tel: 540-999-3500.

Virginia Tourism Corporation

901 East Byrd Street, Richmond, VA 23219; tel: 804-786-2051.

CAMPING

Campgrounds are available on a first-come, first-served basis at Matthews Arm, Lewis Mountain, and Loft Mountain. Big Meadows campground requires reservations in spring and fall; to reserve a site, call 800-365-2267. Campgrounds accommodate tents and trailers; there are no hookups. Permits, required for backcountry camping in the park, are available at visitor centers, entrance stations, and park headquarters.

LODGING

PRICE GUIDE – double occupancy

$ = up to $49 $$ = $50–$99
$$$ = $100–$149 $$$$ = $150+

Big Meadows Lodge

Aramark, Shenandoah National Park, P.O. Box 727, Luray, VA 22835; tel: 800-778-2851 or 540-999-2221.

This cut-stone lodge, built in 1939 and named after a nearby meadow, has an oak-and-chestnut interior. One of two park lodges, Big Meadows offers 20 rooms in the main structure and 72 rooms in rustic cabins and multi-unit lodges with modern suites. Rooms afford panoramic views of the Shenandoah Valley. The restaurant prepares box lunches upon request. Amenities include a playground and ranger programs. The lodge is open early May through late October. $$

The Cabins at Brookside

2978 U.S. Highway 211 East, Luray, VA 22835; tel: 800-299-2655 or 540-743-5698.

Situated west of the park's central region, near Luray Caverns, this stopping place has nine log cabins and a restaurant. Cabins have fireplaces, decks, refrigerators, microwaves, and kitchens; some cabins have two bedrooms, and honeymoon suites have hot tubs and whirlpools. An art gallery and gift shop are on the premises. $$–$$$$

Iris Inn

191 Chinquapin Drive, Waynesboro, VA 22980; tel: 540-943-1991.

The inn is situated on 21 wooded acres at the southern end of the park. Each of the inn's nine rooms has a private bath, refrigerator, pine furnishings, and king- or queen-sized bed. The Great Room, a common area, has a 28-foot-wide stone fireplace and a mural that depicts the wildlife of the Blue Ridge Mountains. A full breakfast is served. $$–$$$

Silver Thatch Inn

3001 Hollymead Drive, Charlottesville, VA 22911; tel: 804-978-4686.

One of central Virginia's oldest buildings, dating to 1780, was built by Hessian soldiers captured during the Revolutionary War. The original two-story log cabin was incorporated into larger structures in 1812 and 1937. The Silver Hatch has seven guest rooms, including a cottage. Rooms have private baths; some have fireplaces and canopy beds. A restaurant and bar are on the premises. $$$–$$$$

Skyland Resort

Aramark, Shenandoah National Park, P.O. Box 727, Luray, VA 22835; tel: 800-778-2851 or 540-999-2221.

This park lodge, originally the summer retreat of George Freeman Pollock Jr., whose efforts led to the creation of the park, was established in 1894. Perched at 3,680 feet (the highest point along Skyline Drive), the lodge offers spectacular views of the Shenandoah Valley. The resort's 177 rooms include rustic cabins and suites, a stable, playground, craft shop, and restaurant. Ranger programs are available. The lodge is open late March to early December. $$–$$$

TOURS

Luray Caverns

U.S. Route 211, Luray, VA 22835; tel: 540-743-6551.

Commercially owned, Luray Caverns, the largest in the state, offers one-hour tours every 20 minutes. One room, 500 feet by 300 feet, has a 140-foot ceiling.

Shenandoah National Park

3655 U.S. Highway 211 East, Luray, VA 22835; tel: 540-999-3500.

Information about the park's guided nature tours is available at visitor centers. Self-guided walks set out from Dickey Ridge, Skyland, Big Meadows, Lewis Mountain, and Loft Mountain. The park encompasses about 500 miles of trails, including 100 miles of the Appalachian Trail.

Skyline Caverns

U.S. Highway 340, Front Royal, VA 22630; tel: 540-635-4545.

Calcite "cave flowers" and a subterranean waterfall are among the attractions of this privately owned cave. Call for a tour schedule.

Excursions

Breaks Interstate Park
P.O. Box 100, Breaks, VA 24607; tel: 540-865-4413.

"The Grand Canyon of the South," a 1,600-foot-deep gorge, lies on the Virginia-Kentucky border. The Russell Fork River carved the canyon over a period of some 250 million years, leaving steep walls towering over a series of low falls and rapids. The park has 12 miles of trails with four scenic overlooks. Whitewater rafting through the canyon is popular.

Delaware Water Gap National Recreation Area
River Road, Bushkill, PA 18324; tel: 570-588-2451

At the border of northwest New Jersey and eastern Pennsylvania, the Delaware River has cut through the 1,500-foot-high Kittatinny Ridge, a wrinkle in the ancient Appalachian range. Hard quartzite composes the ridge, while softer shales and limestones are eroded into flanking valleys. Interstate 80 passes through the gap, but don't overlook the park's quiet back roads and hiking trails. Also consider experiencing the gap the way it began: at river level. This stretch of the Delaware River is popular with novice canoeists.

Great Smoky Mountains National Park
107 Park Headquarters Road, Gatlinburg, TN 37738; tel: 423-436-1200.

Heavy deformation at the southern end of the Appalachians has created a park full of metamorphic rock, extensive faulting, and complex geology. Clingmans Dome (6,643 feet) is a bald knob with a superb view. Look for exposed quartzite in the parking lot. At Cades Cove, a thrust fault has shoved Precambrian mountains over younger rocks, which are exposed in the cove's floor.

Mammoth Cave
National Park
Kentucky

S et in southern Kentucky about 30 miles east of Bowling Green, **Mammoth Cave National Park** encompasses the world's longest known system of underground caverns, with more than 350 miles of surveyed passageways, about 12 miles of which are open to the public. The caves were discovered 2,000 to 4,000 years ago by archaic Indians who used cane torches to search for medicinal minerals such as gypsum, epsomite, and selenite along miles of passageways. Several well-preserved bodies from the period have been found in the cave, including one of a man who had been pinned beneath a six-ton boulder. Much more recently, Mammoth Cave was mined for saltpeter to make gunpowder during the War of 1812, and the remains of the mining works can be seen on some of the cave tours.   ◆   The geology of **Explore a bizarre but beautiful** Mammoth Cave is fairly straightforward. **world of domes, pits, and** Under an eroded caprock of sandstone **shimmering crystals in the** are nearly flat strata of Mississippian lime- **world's longest cave system.** stone that has been slowly infiltrated by groundwater. The erosive force of the water is enhanced by its absorption of carbon dioxide from the air and soil, resulting in a weak carbonic acid that dissolves the calcite (calcium carbonate) in limestone and gradually hollows out the caverns.   ◆   The limestone bed, which originated on the floor of a shallow continental sea about 350 million years ago, has been cut at five separate levels, each corresponding to a different period of the Green River's development. Today the subterranean Echo River flows through the fifth, or basement, level of the cave system. It ultimately drains into the Green River, which now runs on the surface of the park. Although most of the water that created the cave system comes from

Groundwater made acidic by carbon dioxide has dissolved miles of subterranean passages through the limestones of south-central Kentucky.

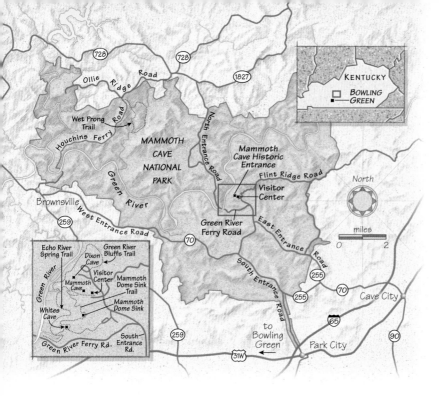

KENTUCKY

□ BOWLING
■ GREEN

728

728

1827

Ollie Ridge Road

Wet Prong Trail

Houchins Ferry Road

MAMMOTH
CAVE
NATIONAL
PARK

North Entrance Road

Mammoth
Cave Historic
Entrance

Flint Ridge Road

Visitor
Center

North

Green River

Brownsville

West Entrance Road

259

Green River
Ferry Road

East Entrance Road

miles

0   2

70

South Entrance Road

255

255

70

Cave City

Echo River
Spring Trail

Green River
Bluffs Trail

Dixon
Cave

Visitor
Center

Mammoth
Cave

Mammoth
Dome Sink
Trail

Green River

Whites
Cave

Mammoth
Dome Sink

259

65

South
Entrance
Rd.

to
Bowling
Green

Park City

90

Green River Ferry Rd.

31W

helmets, lights, and kneepads and spend about six hours climbing, crawling, and sometimes wriggling through undeveloped portions of the cave. Participants must have a chest size no larger than 42 inches; anyone larger will get stuck in the tightest passages. Needless to say, the tour is inadvisable for those who suffer even the slightest whiff of claustrophobia, though most people find it a thrilling and oddly intimate experience with the Earth.

Tour-goers see many classic cave structures, from vertical shafts, which form along fractures in the rock, to enormous hollowed-out chambers and tubular passages, essentially horizontal tunnels. The most beautiful features, however, are the formations created by the slow deposition of minerals carried in the groundwater as it seeps through the rock: travertine flowstones such as the 75-foot-high calcite drapery called Frozen Niagara, gypsum "flowers," and mirabilite crystals, which look like clear needles.

the immediate vicinity, scientists have used dye tracers to determine that some water percolates in from as far away as 20 miles. This has raised concern about the effects of environmental pollutants (from livestock, home septic systems, and agricultural fertilizers) on the caves.

Choose Your Tour

Park visitors usually devote most of their time to one or more of the underground tours. These last from half an hour to the better part of a day; the menu of available tours changes seasonally. Those who take the two-hour **Historic Tour** near park headquarters will see artifacts left by American Indians, see ruins of the saltpeter mines, and observe evidence of Mammoth Cave's early explorers. This was traditionally the most accessible portion of the caverns.

The two-hour **Frozen Niagara Tour** leads to huge pits, domes, and bizarre dripstone formations, including stalactites and stalagmites. On the **Wild Cave Tour**, which covers nearly six miles, visitors are provided with

Surface Features

In many ways, the surface geology is as compelling as that below. Certainly it demonstrates the connection between what happens above and beneath the ground. The park encompasses a fine example of "karst" topography, a term geologists use to describe rugged limestone terrain with a noticeable lack of surface water despite abundant rainfall. Because the underlying limestone is so porous, rainwater quickly seeps into the ground. In some cases,

into the caverns.

Several hiking trails in the park provide informative tours of the region's karst topography. Three of the best trails are in the immediate vicinity of the Historic Cave Entrance. The two-mile **Mammoth Dome Sink Trail** leads to a prominent sinkhole in the forest; the **Echo River Spring Trail**, about half a mile long, descends a hillside to a sizable natural spring; the 1.1-mile **Green River Bluffs Trail** follows the limestone cliffs above the deeply cut Green River. More extensive hiking trails are found north of the Green River. One of the most popular is the **Wet Prong Trail**, which runs for about 4.9 miles in a wandering circuit east of Houchins Ferry Road, offering good views of sandstone ridges and limestone valleys.

it flows into sinkholes (large natural drains that form above cracks in the limestone) or disappearing streams (where surface water simply drops away through a fissure in the rock). Springs gush from rock walls, and brawling brooks disappear just as abruptly. Most of the ridges in the area are capped with sandstone, and most of the valleys are floored with limestone. Rainwater flows off the ridges into the valleys, where it soon finds its way

A caver (above) explores an underground stream.

A waterfall (left) at the Historic Entrance continues the erosional process that created Mammoth Cave.

The pipistrel bat (below) is one of three bat species living at the park.

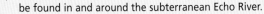

Life in the Underground

The ecology of caves is similar to that of the deep sea: Both are lightless, chilled, inhospitable realms where food energy is scarce and only the simplest life forms can endure. In the perpetually dark and cool chambers of Mammoth Cave, a variety of creatures manage to find small, fragile niches in which to survive and reproduce. These cave dwellers include blind crayfish, spiders, beetles, snails, shrimp, and even fish, most of which can be found in and around the subterranean Echo River.

The cave creatures, or troglodytes, live much as their counterparts do aboveground but without eyes or pigment, both of which are unnecessary in the perpetual night of their subterranean homes. To compensate, some of their other senses – sensitivity to vibrations, for example – are thought to be more acute.

A number of terrestrial animals – bats (three species), woodrats, mice, snakes, and turtles – find seasonal shelter inside the cave entrances. One of the major conservation projects in the park is restoring the population of endangered gray bats to its historic numbers. The installation of baffled gates at various man-made or enlarged entrances will permit bats to enter, but still reduce air flow, which brings in dust, fungus, and outside temperatures.

TRAVEL TIPS

DETAILS

When to Go

Beneath the surface, Mammoth Cave has a year-round temperature of 54°F and high humidity, so a light jacket or sweatshirt is a good idea. Sneakers or hiking boots are highly recommended since most cave trails are damp. Summer highs above ground reach into the 90s; in winter the highs average in the 40s. Spring is cool and rainy; fall is generally pleasant. The park is most crowded June through August.

How to Get There

Commercial airlines serve Nashville, Tennessee, and Louisville, Kentucky, both about 95 miles from the park via Interstate 65.

Getting Around

A car is the most convenient means of travel. Rentals are available at the airports.

Handicapped Access

The visitor center is accessible. The Heritage Trail (on the surface) has been specially designed to accommodate visitors with disabilities. One cave tour is designed for wheelchair use.

INFORMATION

Mammoth Cave National Park

P.O. Box 7, Mammoth Cave, KY 42259; tel: 270-758-2328.

Cave City Convention Center

502 Mammoth Cave Street, Cave City, KY 42127; tel: 800-346-8908 or 270-773-3131.

CAMPING

There are three campgrounds in the park. The Headquarters campground has 111 primitive sites, Houchins Ferry has 12 primitive sites, and Maple Springs has seven group sites that also accommodate horses. To reserve space, call 800-967-2283. Backcountry camping is allowed with a permit; inquire at the visitor center.

LODGING

PRICE GUIDE – double occupancy

$ = up to $49 $$ = $50–$99
$$$ = $100–$149 $$$$ = $150+

Caveland Motel

415 North Dixie Highway, Cave City, KY 42127; tel: 877-773-2321 (toll free) or 270-773-2321.

Fourteen ground-floor rooms offer basic but comfortable accommodations. Amenities include an in-ground pool. $–$$

Cave Spring Farm Bed-and-Breakfast

P.O. Box 365, Smiths Grove, Kentucky 42171; tel: 270-563-6941.

Built in 1857 and now part of a 17-acre estate about six miles from Mammoth Cave, this bed-and-breakfast has 13 rooms, and a one-room schoolhouse from the Civil War period that's now used as a cabin. It also has its own cave, with free daily tours for guests. $$

Mammoth Cave Hotel

National Park Concessions, P.O. Box 27, Mammoth Cave Hotel, Mammoth Cave, KY 42259; tel: 270-758-2225.

This hotel, set in the park adjacent to the visitor center, has 110 rooms ranging in style from modern to rustic. Two rooms accommodate people with disabilities. There are also 38 rustic cottages, a laundry, and camp store. Some rooms have air conditioning and phones. $–$$

Oakes Motel and Campground

5091 Mammoth Cave Road, Cave City, KY 42127; tel: 270-773-4740.

The Oakes has 10 cabins set in an oak grove. Built in the 1930s and refurbished in 1999, the cabins range in size from three rooms (including a kitchen) for up to six adults, to single-room cabins for couples. The motel is 10 minutes from the Mammoth Cave visitor center. $–$$

Rose Manor Bed-and-Breakfast

204 Duke Street, Cave City, KY 42127; tel: 888-806-7673 or 270-773-4402.

This Victorian house has five guest rooms with modern amenities. Horse-carriage rides are available for an extra charge, weather permitting. $$

Wayfarer Bed-and-Breakfast

1240 Old Mammoth Cave Road, Cave City, KY 42127; tel: 270-773-3366.

Built in 1933 as a souvenir shop and refurbished as a bed-and-breakfast, the Wayfarer has five guest rooms with a combination of contemporary and antique furnishings. $$

Wigwam Village #2

601 North Dixie Highway, Cave City, KY 42127; tel: 270-773-3381.

Lovers of wayside kitsch will get a kick out of this motel, built in 1937 as part of a chain of motels with a Wild West theme. Fifteen units, fashioned to resemble teepees, surround the main building. Open March through November, the motel's rooms are air-conditioned but have no phones. $–$$

TOURS

Mammoth Cave National Park

P.O. Box 7, Mammoth Cave, KY 42259; tel: 270-758-2328.

The park offers a rotating menu of tours. The 45-minute Discovery Tour is the shortest. Others last up to six hours and include basic caving instruction (helmets and lights provided). Reservations are strongly recommended; call 270-758-2328. Be prepared to climb stairs and some steep inclines; strollers and camera tripods are not permitted.

MUSEUMS

Crystal Onyx Cave

8709 Happy Valley Road, Cave City, KY 42127; tel: 270-773-2359.

Discovered in 1960 in Pruitt's Knob, Crystal Onyx Cave has many beautiful formations – rimstones, "cave bacon," draperies, stalactites, and stalagmites. Geologists are still exploring how deep the cave goes, and visitors can view a working archaeological dig. Guided tours begin every 45 minutes. The cave also contains a prehistoric Indian burial dated at about 700 B.C. A campground is on the property.

American Cave Museum and Hidden River Cave

119 East Main Street, Horse Cave, KY 42749; tel: 270-786-1466.

The entrance to Hidden River Cave is on Main Street in downtown Horse Cave. The underground chambers were once mined for onyx, saltpeter, and bat guano, and they were the source of the town's drinking water until the 1940s, when pollution became a problem. A museum on the surface sheds light on the cave's geology and history. Hour-long guided tours are available.

Excursions

Red River Gorge Geological Area

Daniel Boone National Forest, Stanton District, 705 West College Avenue, Stanton, KY 40380; tel: 606-663-2852.

Situated about 60 miles east of Lexington, Kentucky, the Red River Gorge contains more than 80 natural arches and bridges, more than any other location east of Arches National Park in Utah. A layer of 320-million-year-old sandstone in the region forms a caprock over softer shale and limestone. About 40 million years ago, tectonic movements bent the sedimentary layers into a broad arch, uplifting the region and cracking open the caprock. This let erosion start to carve the gorge, which is now a spectacular site for hiking (nearly three dozen trails), rock climbing, and whitewater paddling. Nearby Natural Bridge State Park (606-663-2214) has 18 miles of hiking trails and one major centerpiece, the Natural Bridge that lends its name to the park.

Big South Fork National River and Recreation Area

4564 Leatherwood Road, Oneida, TN 37841; tel: 423-569-9778

Big South Fork lies in a remote pocket of the Cumberland Plateau, a densely forested shelf of high country (consisting of a hard layer of sandstone with softer rock beneath) furrowed by hundreds of streams and rivers that flow off the western flank of the Appalachian Mountains. The recreation area has at least nine arches and natural bridges. The best known is Twin Arches, a pair of stone portals eroded side by side into a sandstone ridge. The park is laced with more than 300 miles of hiking trails and is popular with whitewater paddlers.

New River Gorge

P.O. Box 246, Glen Jean, WV 25846; tel: 304-465-0508.

The "Grand Canyon of the East" is a mile wide and about 1,400 feet deep, cut by the poorly named New River, which is actually more than 65 million years old. Once a powerful tributary of the Mississippi, the river was diverted by Pleistocene ice sheets. Before that, the New River had carved its course into the Appalachians as the mountains rose around it. The result is a sheer-walled, gracefully curved canyon exposing rocks as old as 330 million years, including some coal seams. An auto tour from Canyon Rim Visitor Center winds through the area to numerous overlooks and trailheads.

Blanchard Springs Caverns

Arkansas

When you view the Ozark Mountains of northwestern Arkansas, do you see the peaks or the spaces between? No, this isn't some strange twist on a Zen koan. The issue is more geological than philosophical. Unlike the Rockies or Appalachians, whose peaks were uplifted above the surrounding plains, the Ozarks rose up as broad plateaus from which deep chasms or hollows were gouged by millions of years of erosion. In an odd geological twist, it's the valleys rather than the peaks that gave rise to these mountains. ◆ The rocks that form the uplands are relatively soft limestones and sandstones, formed in a warm, shallow ocean more than 350 million years ago. The limestone came from the shells and body parts of billions of marine animals. These settled to the bottom and over the ages compacted into rock.

Water has sculpted both the Ozark Mountains and the pristine caverns deep inside them.

Periodic deposits of sand and mud left layers of sandstone and shale. The horizontal strata can be seen clearly at road cuts and on streamside bluffs throughout the Ozarks. ◆ The same erosive forces that sculpted this spectacular landscape have been relentlessly at work below the surface, too. Like many limestone areas, the Ozarks are abundantly underlain by caves, ranging from shallow setbacks at the base of bluffs (many used as shelters by Native Americans) to long subterranean passageways cut by underground rivers. ◆ Happily, one of the most spectacular of these caves is protected by the U.S. Forest Service just northwest of the small town of **Mountain View** in **Ozark National Forest**. At **Blanchard Springs Caverns**, you'll find not only striking and diverse formations surpassing those of many better-known caves, but also "living"

Draperies of rock form over many thousands of years as groundwater makes countless tiny deposits of calcite.

to
Mountain
Home

ARKANSAS
MURFREESBORO
CRATER OF
DIAMONDS
STATE PARK

Calico Rock

North

miles

0   4

OZARK

NATIONAL

FOREST

BUFFALO
NATIONAL
RIVER

Buffalo
River

White River

Sylamore
Scenic
Byway

North
Sylamore
Creek   Blanchard
Sylamore   Springs
Trail   Caverns
Visitor
Center

Blanchard
Springs

Sylamore
Scenic Byway

Mountain
View

Sylamore

White   River

to
Batesville

Ozark   Mountains

found that someone
had long preceded
their efforts, with fatal
results. A skull, a few
bones, and a burnt-reed
torch told the mute
story of an ancient
Indian visitor – but it's
a mystery how and
why he entered. Carbon
dating put the remains
at A.D. 893.

The U.S. Forest
Service began develop-
ing Blanchard Springs
Caverns in the mid-
1960s, studying other
caves across the country
to learn what to do,
and what to avoid,
before starting tours in

caverns – that is, caves still in the process
of growing – offering visitors a lesson in
how such underground wonders take shape.

A "New" Cave

The natural beauty of Blanchard Springs
Caverns is due in large part to the cave's
relatively recent discovery. Not until 1934 did
a U.S. Forest Service worker descend the deep
hole known as Half Mile Cave, one of the
caverns' natural entrances, and it wasn't until
the mid-1950s that cavers, or spelunkers,
began serious exploration. In 1955, surveyors

Small streams (left)
continue the work of
erosion, channeling
water underground
and keeping the
caverns "alive."

Dripstone Trail
(opposite, bottom)
leads visitors past
spectacular calcite
formations.

Hawksbill Crag
(opposite, top),
undercut by erosion,
juts into space.

1973. Airlocks at the entrances, for example,
keep dry outside air from penetrating the
cave, which has a humidity of nearly 100
percent and a year-round temperature of 58°F.
The Forest Service is still taking steps to
minimize human impact on the cave environment. The trails are carefully designed and
maintained, the lighting is subdued, and
there are limits on the size of tour groups.
Unlike many privately owned caves, you'll
find almost no broken or vandalized formations and none of the garish lighting or
painted surfaces that are supposed to
appeal to tourists. Nor does one find the
grandiose names that proliferate elsewhere.
What's more, the cave sustains most of its
original inhabitants, from invertebrates such
as crickets and crayfish to larger animals
such as salamanders and bats.

Going Underground

The road to the caverns, part of the Ozark
National Forest's **Sylamore Scenic Byway**,
winds through oak-hickory woodland to a
visitor center with a small museum and book-
shop. In an adjoining theater, rangers show
a short film explaining the geology of the
Ozarks and the formation of the cave system.

Two tour routes are offered at the caverns;

both begin with a 216-foot elevator descent beneath the surface. The most popular with first-time visitors is the half-mile **Dripstone Trail**, which passes through two large chambers in the higher and older portion of the cave.

The strata along the trail consist of layers of limestone from the upper Silurian and lower Devonian periods that are roughly 400 million years old. Most beds of this type average three feet thick, but the Sylamore shale and sandstone reach 15 feet. The strata are capped on top by the Boone Formation, a bed of chert 250 feet thick that is relatively impervious to groundwater. A similarly resistant layer, the 150-foot-thick St. Peter sandstone, sandwiches the caverns from below.

The geological processes that created the cave formations, or speleothems, along the trail have had hundreds of thousands

of years to do their work. You'll find a dazzling variety of stalactites, stalagmites, flowstones, draperies, "popcorn," and other configurations. All were formed by the gradual accretion of calcite crystals (calcium carbonate) that were dissolved by seeping water out of the limestone above and redeposited drop by drop. In many places, the calcite is almost pure and looks like glistening white crystals. But where water has picked up trace minerals such as iron or manganese, formations are

tinted tan, rust, gray, or black. The painstaking work of cave building continues today. The faint sound of dripping water is heard just about everywhere.

Lucky visitors may spot a cave cricket (actually a small grasshopper) or a four-inch-long Ozark blind salamander, and, at certain times of the year, inch-long juvenile salamanders wriggling in trailside pools. These creatures are small and shy, however, and quite difficult to spot. Five species of bats, including the endangered Indiana and gray bats, reside in the cave as well, and while these flying mammals are rarely seen, huge mounds of guano testify to their presence.

Steps of Discovery

The longer of the caverns' two tours, the 1.2-mile **Discovery Trail**, is open from Memorial Day through Labor Day. Winter closing helps protect the bats that hibernate in its passageways. The trail traverses nearly 700 steps and is not recommended for those with respiratory

Crater of Diamonds

Folded, faulted, and elevated by an ancient continental collision, southwestern Arkansas' **Ouachita Mountains** are one of the few east/west-oriented ranges in North America. Set among these pine-covered ridges is an even more notable geological site, much more important than its relatively small size indicates. Just south of the little town of **Murfreesboro**, millions of years of erosion have uncovered a 36-acre field of kimberlite, an igneous rock brought up from deep within the Earth by volcanic activity. And scattered within the kimberlite lie tiny fragments of highly compressed pure carbon – better known as diamonds.

After diamonds were discovered here in 1906, several attempts were made to mine the deposit commercially. All failed, however, and today the site is encompassed by **Crater of Diamonds State Park**. For a small fee, visitors are allowed to roam the rocky field – the exposed top of a narrow volcanic "pipe" – and keep any gems they find. About two diamonds a day turn up, and while most are small, low-quality stones, lucky searchers have made some astonishing finds over the years. In 1924 a 40-carat diamond was unearthed here and is still the largest ever found in the United States. In 1990 a woman walked away with a flawless stone of more than three carats, valued after cutting at more than $30,000.

Park officials happily offer diamond-hunting tips for first-time visitors, and they plow the kimberlite field regularly to expose fresh soil. Even if you don't find a gem-quality diamond, you could take home a chunk of jasper, amethyst, or agate – a pretty souvenir of your visit to one of North America's most unusual geological phenomena.

Natural bridges (above) occur as erosion opens joints in the rock and weaker pieces fall away.

Limestone cliffs and ledges (opposite) are characteristic of the Ozarks.

Small diamonds (left) are routinely found at Crater of Diamonds State Park, where rangers encourage visitors to prospect for their own (below).

or other health problems.

Deeper than the Dripstone Trail, the Discovery Trail takes you to caverns that are far younger than their upstairs neighbors, although, para-doxically, the surrounding rock, a 120-foot-thick slab of Plattin limestone laid down in the Ordovician Period, is much older.

Because of the chambers' relative youth, speleothems have had less time to grow, and therefore you'll see fewer elaborate formations. You can walk alongside the sort of underground stream that cre-ated these caverns, however, and see fresh evidence of its work. Smooth-sided tunnels testify to the sculpting power of moving water. The trail also passes beneath the entrance once known as Half Mile Cave. The first surveyors entered here, descending on ropes, and signs of their campsites can still be seen. The bones of that first, ill-fated Indian explorer were found nearby.

This is not to say that the Discovery Trail is lacking in splendor. Both stalagmites and stalactites are present; the latter are especially evident in a spot called **Stalactiflats**. Delicate-looking rimstone dams and terraces have formed along the stream, and a gleaming, white formation called the **Giant Flowstone** near the end of the trail is an astonishing sight, exquisitely reflected in the still water of the pool below. All in all, anyone interested in caves or Ozarks geology should see both trails to understand this fascinating subterranean environment.

Between tours, a drive downhill from the visitor center brings you to a boardwalk trail through a little hollow where the caverns' underground river pours out of **Blanchard**

Springs. Flowing from a room-sized hole in a cliff at a rate of about 7,000 gallons per minute, the spring is joined by dozens of rills before flowing into beautiful **North Sylamore Creek**, one of the clear, rocky streams typical of the Ozarks. Like many others, it's a pop-ular summertime swimming hole for local residents and visitors staying at the nearby Forest Service campground. The limestone bluffs rising above the creek offer evidence of the ancient origins of the mountains. But on a hot summer day, geological theory is likely to give way to the simple delight of a plunge into the cool, clear water.

TRAVEL TIPS

DETAILS

When to Go

The temperature in the cave is 58°F year-round, with relative humidity near 100 percent. Footgear with nonskid soles is highly recommended as the trails underground are wet in most places. In general, the Ozark Plateau has warm, humid summers, with average daytime highs in the upper 80s. Colorful fall foliage attracts many visitors to the region.

How to Get There

Major airlines fly into Little Rock, 110 miles south of Mountain View. Small aircraft serve Harrison, 80 miles from Mountain View.

Getting Around

A car is the most convenient means of travel. Rentals are available at the airports.

Handicapped Access

The visitor center is fully accessible. The Dripstone Trail is wheelchair-accessible; assistance is needed on some steep ramps.

INFORMATION

Blanchard Springs Caverns

P.O. Box 1279, Mountain View, AR 72560; tel: 870-757-2211 or 888-757-2246.

Mountain View Area Chamber of Commerce

P.O. Box 133, Mountain View, AR 72560; tel: 870-269-8098 or 888-679-2859.

CAMPING

Blanchard Springs Campground

Sylamore Ranger District, P.O. Box 1279, Mountain View, AR 72560; tel: 870-757-2211 or 888-757-2246.

This campground near the caverns' entrance has 32 sites, flush toilets, and hot showers. Sites are available on a first-come, first-served basis.

Gunner Pool Campground

Sylamore Ranger District, P.O. Box 1279, Mountain View, AR 72560; tel: 870-757-2211 or 888-757-2246.

About 10 miles from the caverns, the Gunner Pool campground has 27 sites also available on a first-come, first-served basis.

LODGING

PRICE GUIDE – double occupancy

$ = up to $49 $$ = $50–$99
$$$ = $100–$149 $$$$ = $150+

Best Western Fiddler's Inn

HC 72, Box 45, Mountain View, AR 72560; tel: 870-269-2828 or 800-528-1234.

This 48-room, chain motel lies half a mile north of the junction of highways 5, 9, and 14 in Mountain View. It has a pool, two-room suites, and non-smoking rooms. A restaurant is nearby, and meals can be delivered. $–$$

Dogwood Motel

HC 71, Box 86, Mountain View, AR 72560; tel: 870-269-3847 or 888-686-9275.

On Highway 14 a mile east of downtown Mountain View, this 30-room motel is open year-round and offers basic comfort and a pool. $–$$

Dry Creek Lodge at Ozark Folk Center State Park

P.O. Box 500, Mountain View, AR 72560; tel: 870-269-3871 or 800-264-3655.

This motor inn is operated by the state Department of Parks and Tourism and is the venue for a folk music festival held on the third weekend of April. It has 60 rooms with private patios, a pool, and a wooded setting. A restaurant is nearby. $$

Inn at Mountain View

307 West Washington Street, Mountain View, AR 72560; tel: 870-269-4200.

This Victorian bed-and-breakfast, built in 1886, offers 10 guest rooms with high ceilings, carved woodwork, and antique furnishings. A full breakfast is included. $$

Wildflower Bed-and-Breakfast

100 Washington Street, Mountain View, AR 72560; tel: 870-269-4383.

Set on the town square in Mountain View, this bed-and-breakfast was originally built in 1918 as the Commercial Hotel. It now has five single rooms and three suites, each with a private bath. $$

TOURS

Blanchard Springs Caverns

P.O. Box 1279, Mountain View, AR 72560; tel: 870-757-2211 or 888-757-2246.

The caverns offer two underground tours for the general public: the Dripstone Trail and the Discovery Trail. A "wild cave" tour designed to give novices a taste of caving is also available by reservation; contact the office for details.

Bull Shoals Caverns

P.O. Box 444, Bull Shoals, AR 72619; tel: 870-445-7177.

The caverns are located near the Arkansas–Missouri border and are known for interesting calcite formations, including boxwork and cave pearls. Guided tours are given every half hour.

Mystic Caverns

P.O. Box 13, Dogpatch, AR 72648; tel: 870-743-1739.

Located eight miles south of Harrison, Arkansas, on Highway 7, Mystic Caverns has two caves on show–Mystic Cavern itself, which was discovered in the 1850s and has been open to visitors since the 1920s, and Crystal Dome Cavern, found in 1968 and open since 1981. The latter has an eight-story-high "dome," and both have many interesting calcite features such as flow-stones, rimstones, and soda straws. Hour-long guided tours are available.

Ozark Ecotours

P.O. Box 513, Jasper, AR 72641; tel: 870-446-5898.

Canoeing is popular on the White and Buffalo Rivers, which border the forest district to the north, east, and west. Ozark Ecotours runs floating and hiking trips on the Buffalo River and tours into several other caves in the region.

Excursions

Hot Springs National Park

P.O. Box 1860, Hot Springs, AR 71902; tel: 501-624-3383.

Groundwater heated geothermally to 143°F issues from 320-million-year-old sandstone at the base of Hot Springs Mountain, part of the ZigZag Range of the Ouachita Mountains. The water contains silica, calcium, and other minerals, and has been used for healing since prehistoric times. Numerous hiking trails and roads wind through the park in the highlands around the bathhouses. The area is also known for quartz crystals, found along the ridges, and for novaculite (also known as "Arkansas stone"), a mixture of quartz and calcedony. There is a good exposure of white novaculite at the turnaround atop West Mountain.

Ouachita Mountains

Ouachita National Forest, P.O. Box 1270, Hot Springs, AR 71902; tel: 501-321-5202.

The folded and faulted Ouachita Mountains push up west-central Arkansas like a bump of ancient debris beneath a rug. The rocks are primarily Paleozoic sandstones, shales, novaculites, and cherts. The novaculites form the southern ridges, cherts and sand-stones the northern. Take the Crystal Vista Auto Tour to Crystal Mountain, where you can collect mineral specimens. Or check with the U.S. Army Corps of Engineers (501-767-2108) about the Lake Ouachita Geo Float, a waterborne field trip.

New Madrid

New Madrid Historical Museum, 1 Main Street, New Madrid, MO 63869; tel: 573-748-5944.

A series of massive earthquakes in 1811 and 1812 violently shook the central Mississippi valley near this town in the southeast corner of Missouri. Buildings collapsed, waves a couple of feet high coursed through the earth, and nearby Reelfoot Lake in Tennessee was created. The biggest quake was so powerful that it woke President James Madison in Washington, D.C., and rang church bells in Boston. Begin exploring the region at the New Madrid Historical Museum, then cross into Tennessee and see Reelfoot Lake near Tiptonville.

Badlands National Park
South Dakota

CHAPTER 9

The turnoff for **Badlands National Park** is at Exit 131 on Interstate 90, some 70 miles east of **Rapid City**, South Dakota. And for the first few miles it's the same scenery you've been looking at ever since crossing the Missouri River: vast expanses of grassy fields and pastures dotted with cattle, farmhouses, windmills, and silos. This is farm and ranch country, formerly buffalo and Indian country, unmistakably Western, tufted with yucca and wheat grass, domed with a soaring sky. ◆ Then, with startling suddenness, as you travel south on **Loop Road 240**, the landscape is transformed. The ground falls away, the grass disappears, and the land buckles and splits. A fantastic array of spires, turrets, and pinnacles unfolds like a pop-up in a children's book. You've entered Badlands National Park, where the weathered land magically opens and exposes the strata underpinning the terrain. ◆ The Loop Road follows the **Badlands Wall** for 25 miles, running west from the visitors center at **Cedar Pass** to the turn-off for the town of **Wall** on I-90, north of the park.

A jumbled landscape of sawtooth cliffs, gnarled spires, and twisted canyons has been torn from the prairie by half a million years of erosion.

(Stop at the visitors center for a guide brochure and interesting exhibits.) The road winds along the base of the Badlands Wall before ascending to the crest at **Norbeck Pass**, then dips from Dillon Pass, and rises again to the **Pinnacles Overlook**. ◆ The view from the top of the Badlands Wall is dazzling. To the north, undulating grassland runs to the horizon. To the south, the land drops away several hundred feet in a jumble of eroded slopes and cliffs, a stupendous vista that takes in the chalky swath of the White River and the north edge of the Pine Ridge Indian Reservation. It's like standing at

Clouds gather over South Dakota's Badlands; the soft shales and siltstones are easily eroded by the infrequent but forceful downpours.

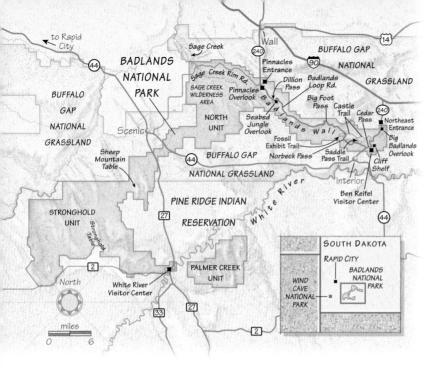

here are soft and riddled with cracks and holes. Rocks crumble underfoot; you can squeeze some of them to powder. Water trickling through the surface generates a kind of internal bleeding that creates slumps such as **Cliff Shelf**, which slid down the face of the Wall near the top of Cedar Pass.

It all happens swiftly. The colorful haystack mounds along **Dillon Pass** are disappearing at a rate of an inch a year. Between 1915 and 1950, **Vampire Peak** lost almost 20 feet.

Building and Carving

Four major geologic formations are visible in Badlands National Park. From bottom (oldest) to top, they are the Pierre shale and the Chadron, Brule, and Sharps Formations. Stacked on one another like carpet samples on a counter, each is identifiable by color and texture.

Eighty million years ago, during the late Cretaceous Period, this portion of Earth was covered by a shallow inland sea. Sixty-eight million years ago, the ancestral Rocky Mountains to the west began to rise, draining the sea and exposing layers of black mud. The top layers of this deposit, known today as the Pierre shale, underwent a chemical reaction that fossilized the various soils and altered their color to bright yellow and deep lavender.

Next up the geologic ladder is the Chadron Formation. Thirty-seven to 32 million years ago, during the Eocene Epoch, rivers leading north and east from the Black Hills changed from clear, quiet streams to raging torrents choked with sand and gravel. These spread over the land, filling in valleys that had been scoured out of Pierre shale. Over several million years, hundreds of feet were laid down,

the lip of a dry waterfall, looking down a long, slanting cataract of stone.

The wall itself is fragile. Annual rainfall here averages about 15 inches, most of it in summer's short, crackling cloudbursts. Downpours carry off the rock in myriad rivulets toward the White River. The soil dries quickly to a shimmering gloss, inhibiting root and plant growth. With nothing to hold the soil, little prevents the relentless process of erosion.

The sedimentary layers of the Badlands look and feel very different from the rocks of, say, Arizona's Grand Canyon. Composed of claystone, siltstone, and sandstones, the layers

Badlands Stratigraphy

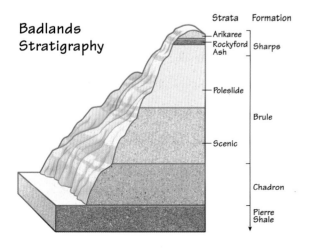

Strata	Formation
Arikaree	
Rockyford Ash	Sharps
Poleslide	
	Brule
Scenic	
	Chadron
Pierre Shale	

The **Notch Trail** (left) leads up a steep ladder and along a narrow ledge to a stunning view of the Badlands.

A **field worker** (below) takes notes at the "Big Pig Dig," site of an *Archaeotherium*, a distant relative of the modern pig.

mentary layers that eroded at different rates. This accounts for the fantastic variety of shapes seen today in the Brule layer.

The Sharps Formation, from the end of the Oligocene Epoch (28 to 26 million years ago), crowns the ridgeline of the tallest peaks along the Badlands Wall. The formation is largely volcanic ash blown from eruptions in Wyoming and Colorado. It blanketed the area and was later covered by stream deposits. Some of the park's steepest cliffs are found in the Sharps – for example, the serrated crest of Pinnacles Overlook.

At the end of the Pliocene Epoch (five million years ago), Badlands deposits stacked up at least 500 feet higher than today's ground level. If sedimentary deposition was steady, its destruction has been relentless.

most of which gradually eroded away, leaving virtually no record of their presence. All that's left is a rounded, pastel layer averaging about 30 feet thick in places such as **Dillon Pass**.

The Brule Formation, deposited in the Oligocene Epoch (32 to 26 million years ago), is recognizable by steep slopes, pinkish horizontal banding, and an ashy texture that resembles a kind of oatmeal. The Brule is rich in fossils, especially turtles, oreodonts (sheeplike creatures), and tiny deer known as Leptomeryx. (Remember, fossil collecting is forbidden in the park.)

The climate back then was mild and moist. Sluggish streams overflowed their banks and shifted course as sandbars and mudbanks choked their channels, drowning animals in periodic floods. The terrain was mostly flat, composed of hard and soft sedi-

Inundation and Erosion

About four and a half million years ago, the Great Plains began to tilt strongly east, which increased the gradient of the streams flowing out of the mountains to the west. The climate cooled. About one million years ago, the Pleistocene ice sheets began to nudge down from the north. While the ice never reached the Badlands, the **Missouri River** was forced to detour south along the flank of the advancing sheet. A mere 500,000 years ago, the young Cheyenne River captured the streams flowing east out of the Black Hills, making them flow northeast to the Missouri, rather than due east to the Badlands. Without the replenishing soils, erosion quickly carved the soft Badlands sediments into their current configurations.

Just about everything in this geological mosaic comes together for the space of a mile or two along Dillon Pass, at the heart of the Loop Road. At **Seabed Jungle Overlook**, down in a ravine at the base of a haystack formation, you can see the silky black soil

of Pierre shale, mounded by bright bands of yellow and red, topped by remnants of Chadron, Brule, and Sharps deposits. It's the only place in the park where all four major formations can be viewed at once.

The Badlands story is one of inundation and erosion. Back and forth for unimaginable years came successive waves of waterborne deposits – inland seas, sluggish rivers, and deltas. Then, as the climate dried and cooled, these were supplanted by faster, quicker streams, smaller forests, and more spacious savannas that supported a variety of small mammals. Each of these inundations left huge deposits of clay, silt, and sand, which compacted under their accumulated weight. Exposed to sunlight and air, their soils transformed chemically, altering in color and composition.

A Slab of the Moon

A million years is but a blip on the geological radar screen, but such a span of time is virtually incomprehensible to today's visitor. Badlands National Park is a kind of chart that helps you grapple with eons of geological shape-shifting. You can observe the strata and make comparisons, recognize trends and detect clues. But to experience this chart firsthand, you have to get out of your car and walk: along **Castle Trail** at the top of the Wall … Up **Saddle Pass Trail**, which winds through the Badlands Wall … Along **Fossil Exhibit Trail**, with its displays of fossilized animal bones… Around the campground at **Sage Creek Wilderness Area**, with its

Wind Cave

Wind Cave in South Dakota is one of the world's longest, oldest, and most complex caves. Its name comes from how it "breathes" air as the atmospheric pressure changes, inhaling as high pressure passes over and exhaling during lows. To date, more than 80 miles of labyrinthine passages have been explored, a mere five percent of the cave's estimated length.

But you can't just throw on your grubbies, grab a caver's hard-hat, and plunge below. Anyone entering the cave must agree to protect its fragile formations by following established routes. Organized groups that obey the stringent requirements may apply for exploration permits. **Wind Cave National Park** has an ongoing relationship with several knowledgeable caving clubs.

The venture isn't for those with claustrophobia. The route frequently passes through tight openings that entail vigorous wiggling. And it's dark down there. Even the best cavers become disoriented in a world where time warps into new units of measurement.

Wind Cave was formed some 350 million years ago, when a warm, shallow sea deposited sediments that later hardened into layers of limestone and dolomite. Water trickling through the strata dissolved part of the rock; the uplift of the Black Hills 60 million years ago enlarged existing cracks and helped make new ones.

Ordinary cave formations, such as stalactites and stalagmites, are rare here. Instead, Wind Cave boasts the world's greatest variety of "boxwork." This delicately colored, honeycomb formation of translucent calcite fins slants at jagged angles from ceilings and walls. The calcite grew in the joints and cracks of the limestone; when the limestone eroded, the tougher calcite veins stood out. They are Wind Cave's signature formation.

Hiking in the Badlands (left) is best after several days of dry weather; rain can turn trails into a sticky gumbo.

Calcite crystals (right) in the cracks of Wind Cave's limestone formed interlocking veins called boxwork.

Sunset paints the Wall (below) with warm light. One early explorer likened the Badlands' eroded landscape to a "magnificent city of the dead."

exposed slabs of Pierre shale shining in sunlight like veins of coal.

To the impressionable eye, the formations at the north rim of Sage Creek Wilderness Area resemble a scabrous, fissured slab of the Moon. The runnel-marked folds of a collapsed shelf near Pinnacles Overlook conjure up a vision of eviscerated intestines. The vertical ribs of rock known as clastic dikes that stretch for miles near **Stronghold Table** recall lines of thread stitched into the fabric of a tattered quilt. The smooth slopes leading down through **Big Foot Pass** evoke the neck and flanks of a slumbering pachyderm. The images are endless.

Provocative landscapes elicit provocative responses. They also operate as catalysts. They pique our fancy, spark our imagination, and arouse a sense of wonder and delight. At Badlands National Park, the cutaway view of the Earth extending back 80 million years offers a spectacle of form and light that joggles our aesthetic sense and leaves us marveling at the way time has worked its magic over the face of the Earth in this special place.

DETAILS

When to Go

Summer is hot and dry, with daytime highs in the low 90s and occasional thunderstorms. Winter is cold and blustery, with highs in the low 30s. Expect high winds in any season. Spring and fall are generally pleasant, with warm days and cool nights. The greatest number of visitors arrive between mid-June and late September, the fewest from mid-November through March.

How to Get There

Commercial airlines serve Rapid City, South Dakota, about 40 miles from the park.

Getting Around

The Badlands region is vast and remote. A car is the only convenient means of travel. Rentals are available at the airport.

Handicapped Access

The visitor centers and most scenic overlooks and wayside exhibits are accessible. The Fossil Exhibit Trail is wheelchair-accessible but steep in places. The Window Trail is fully accessible; parts of the Door Trail are accessible with assistance.

INFORMATION

Badlands National Park

P.O. Box 6, Interior, SD 57750; tel: 605-433-5361.

The Ben Reifel Visitor Center, at Cedar Pass, is open year-round; the White River Visitor Center, on Highway 27 on the Pine Ridge Indian Reservation, is open from May to August.

Wall Chamber of Commerce

503 Main Street, P.O. Box 527, Wall, SD, 57790; tel: 605-279-2665 or 888-852-9255.

CAMPING

Badlands National Park

P.O. Box 6, Interior, SD 57750; tel: 605-433-5361.

The Cedar Pass Campground, near the Ben Reifel Visitor Center, is open year-round and has 96 sites with flush toilets. The primitive Sage Creek Campground has 10 sites with pit toilets and no water. Campsites are available on a first-come, first-served basis and fill up quickly in July and August. Backcountry camping is permitted anywhere in the park at least a half mile from a road or trail. Backcountry permits are not required, but visitors should check in at the visitor center before heading off-trail.

Badlands Interior Campground and Motel

HC 54, Box 115, Interior, SD 57750; tel: 605-433-5335.

This commercial campground has 70 sites, flush toilets, a pool, and coin laundry. It is about two miles from the Ben Reifel Visitor Center.

LODGING

PRICE GUIDE – double occupancy

$ = up to $49 $$ = $50–$99
$$$ = $100–$149 $$$$ = $150+

Badlands Budget Host Motel

HC 54, P.O. Box 115, Interior, SD 57750; tel: 605-433-5335 or 800-388-4643.

The motel is about two miles from the Ben Reifel Visitor Center and has 21 rooms, a pool, play-ground, and convenience store. Closed October through April. $

Best Western H & H El Centro Motel

P.O. Box 37, Kadoka, SD 57543; tel: 605-837-2287 or 800-837-8011.

About 20 miles east of the park, the El Centro has 39 rooms, indoor and outdoor hot tubs, a heated pool, and restaurant. Closed December and January. $–$$

Best Western Plains Motel

712 Glenn Street, Wall, SD 57790; tel: 605-279-2145.

This chain motel offers 74 rooms in the town of Wall, a popular tourist stop about seven miles north of the park. A pool is available. Closed December through February. $–$$$

Cedar Pass Lodge

P.O. Box 5, Interior, SD 57750; tel: 605-433-5460.

The Oglala Sioux tribe operates this park lodge next to the Ben Reifel Visitor Center. It has small cabins with 24 air-conditioned rooms, including three two-bedroom units. Closed November through February. $–$$

Jobgen Ranch Homestead Bed-and-Breakfast

24800 Sage Creek Road, Scenic, SD 57780; tel: 605-993-6201.

The inn is set on a 5,500-acre working ranch about 30 minutes from the park. The two guest rooms are in a detached house with simple but comfortable furnishings. $$

TOURS

Badlands National Park

P.O. Box 6, Interior, SD 57750; tel: 605-433-5361.

Rangers give daily programs at Cedar Pass in the summer and lead nature walks on several trails. Call the visitor center for details.

MUSEUMS

Badlands Petrified Gardens

Off Interstate 90 at exit 152, Kadoka, SD 57543; tel: 605-837-2448.

This museum has an outdoor collection of large pieces of petrified wood and tree stumps. Indoor displays show fluorescent minerals and fossils from the Badlands region.

Mammoth Site

1800 Highway 18 Bypass, P.O. Box 692, Hot Springs, SD 57747; tel: 605-745-6017.

About 26,000 years ago, several mammoths fell into a sinkhole and drowned. When their remains were discovered in 1974, a nonprofit museum was built over the site to shelter the excavations and interpret their significance. Guided tours and exhibits focus on Ice Age animals. Children enjoy a simulated mammoth dig.

National Grasslands Visitor Center

708 Main Street, P.O. Box 425, Wall, SD 57790; tel: 605-279-2125.

The Buffalo Gap National Grassland adjoins Badlands National Park; the visitor center in Wall features exhibits and educational programs that cover the plants, animals, and geology of the area. The displays also discuss the effects of farming and grazing and the role of wetlands in preserving wildlife habitat.

Excursions

The Black Hills

Black Hills National Forest, Box 200, Custer, SD 57730; tel: 605-673-2251.

Like an enormous rock blister, the Black Hills formed when Precambrian granites rose up about 70 million years ago, lifting the overlying layers of Paleozoic limestones and Cretaceous sandstones. These eroded, leaving the igneous core of the hills flanked by tilted layers of sedimentary rock. The national forest is webbed with trails, and scenic Highway 385 traverses the Black Hills north to south.

Devils Tower National Monument

P.O. Box 10, Devils Tower, WY 82714; tel: 307-467-5283.

Devils Tower rises from the prairie like a gigantic tree stump of igneous rock. It is an eroded remnant of a large lava flow, although how large is still under debate. The flutings in the tower are joints that formed when the molten rock cooled and shrank. Take the Tower or Red Beds Trail, and note the gigantic blocks of talus that have tumbled off the sides of this monolith. The rock contains large white crystals and is "phonolitic" – it clangs somewhat musically when struck.

Theodore Roosevelt National Park

P.O. Box 7, Medora, ND 58645; tel: 701-623-4466.

The Badlands of South Dakota have a close relative in western North Dakota, where erosion has sculpted colorful Tertiary layers of clay, siltstone, and sandstone, mixed with beds of lignite or soft coal. The Scenic Drive in the North Unit leads to Oxbow Overlook, where Ice Age glaciers redirected the Little Missouri River to its present course.

Yellowstone National Park

Wyoming

Yellowstone National Park appears serene, but geologists tell us that this is an illusion. The park sprawls across an enormous volcanic caldera whose repeated eruptions have been greater than any known to history. This makes visiting Yellowstone a little like touring a vast disaster area, the ruins partially rebuilt but still smoldering ominously here and there. Belying the quiet mountain beauty, the geysers, hot springs, warm ground, and frequent earthquakes are inescapable signs that the dragon is not dead, just dozing. ◆ To begin the story, we go back 50 million years, to the birth of the **Absaroka Range**. We could start earlier, during the uplift that raised the **Rocky Mountain** region 70 million years ago. But there isn't much of that landscape left to see. Most of it has been blown up or buried. ◆ The Absarokas start north of the park, run down its eastern boundary, and stretch southward. They grew through several alternating cycles of activity and quiescence lasting

The park's geysers and hot springs are the simmering remains of catastrophic eruptions.

10 million years. Layer by layer, lava and ash spilled from volcanic vents, gradually building a plateau more than 100 miles long and 10,000 feet high. ◆ After the eruptions ceased about 40 million years ago, the area experienced a period of relative calm. There were episodes of uplift and erosion, but nothing comparable to the immense force gathering itself in northern Nevada. That force burst onto the scene 16 million years ago. Huge eruptions and outpourings of lava began along the Nevada-Oregon border and moved across southern Idaho toward Yellowstone. The path the eruptions took is today's **Snake River Plain**. This plain is clearly shown on any relief map, where it looks as though a gigantic thumb has smeared

Colored terraces of travertine at Mammoth Hot Springs grow by eight inches a year as calcite dissolved in hot groundwater is deposited on the surface.

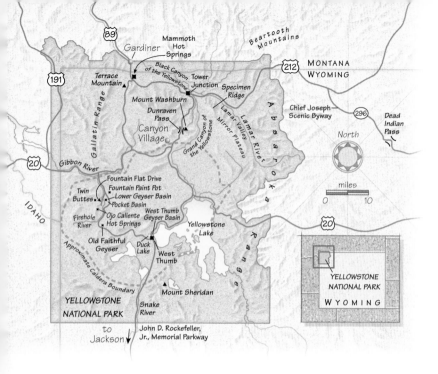

At Yellowstone, this has happened three times in the last two million years. The first and largest created the Huckleberry Ridge caldera 2.1 million years ago. Then came the Island Park caldera just outside the park's western boundary 1.3 million years ago, and finally the Lava Creek caldera 630,000 years ago in south-central Yellowstone. When people speak of the Yellowstone caldera, they usually mean Lava Creek. And when you look at the above numbers, you can see why geologists think that it is getting about time for another eruption – give or take a couple of million years.

a broad, flat track through the mountains.

Which is not far from the truth. The "thumb" is the Yellowstone hot spot, a plume of molten magma extending from the Earth's core to the surface. The plume appears to have migrated northeast. But in reality it remained stationary as the continent drifted slowly over it, like a sheet of steel passing over a blowtorch. Typically, an eruption would last a few days or weeks and then subside to form a giant caldera – the crater of a collapsed volcano.

It's hard to picture the scale of these eruptions since nothing in recorded history comes close. The last eruption ejected something like 240 cubic miles of material. Flows of hot ash sped outward at speeds up to 100 miles per hour; so much ash was blasted into the air that traces can be found around the world, while in Iowa and Kansas

Tower Creek (opposite, top) cuts a channel through Eocene-age volcanic mudflows and plunges over them at Tower Falls.

Steam rises from Grand Prismatic Spring (right) along the aptly named Firehole River in the Midway Geyser Basin

A cross-country skier (opposite, bottom) marvels at Riverside Geyser, also along the Firehole.

Yellowstone ash deposits lie inches thick. Judging from recent eruptions such as Mount St. Helens in 1980 (which ejected less than one cubic mile) and Indonesia's Mount Tambora in 1815, the sound alone may have been audible thousands of miles away.

Heat Beneath the Ice

When the Lava Creek volcano collapsed, its caldera measured 45 by 28 miles and was several thousand feet deep. Subsequent lava flows have obscured the caldera walls and built the hills and bluffs, now covered by lodgepole pine, that occupy the central and southwest parts of the park.

Lava was still oozing 70,000 years ago when the Pinedale glaciation began, the most recent of perhaps a dozen glaciations. The cold reached its peak 25,000 years ago, with an ice cap several thousand feet thick. This was a mountain glacier, distinct from the vast continental ice sheets, which did not reach the park. Centered over **Yellowstone Lake** and augmented by glaciers from surrounding mountain ranges, the Yellowstone ice cap pushed into **Jackson Hole** to the south and **Paradise Valley** to the north. Areas at lower elevation remained ice-free, so when the glaciers retreated 10,000 to 14,000 years ago, animals and plants were ready to recolonize the newly exposed land. Thermal areas – active warm zones – stayed hot beneath the ice and must have created some unusual cavities even during the coldest times.

The hot spot's plume is still there and still powering geothermal displays. Among the heralds of future activity are two resurgent domes, areas within the old Lava Creek caldera that seem to be blistering upward. Major concentrations of geysers and hot springs lie on the periphery of these domes.

Exploring the Caldera

The best place to see the Yellowstone caldera is from the slopes of **Mount Washburn** north of **Canyon Village**. You can hike to the summit or drive to the overlooks on the south side of **Dunraven Pass**. Either way, you are on the edge of the caldera. To appreciate its size, locate **Mount Sheridan**, 40 miles south across **Yellowstone Lake**. Sheridan and Washburn were once points along the same branch of the Absaroka Range. The Yellowstone volcano obliterated

Yellowstone's Geysers

The concept is simple. It's the execution that astonishes and enthralls.

A geyser works by pressure and sudden release. Water from rain and melting snow percolates into the ground, where it gets hot enough to boil, and returns to the surface as a fumarole (steam vent) or a hot spring. If something blocks the upward flow – in effect puts a lid on the system – the water gets hotter and builds up pressure. When the system lets go, the superhot water explodes into steam, the lid is blown off, and there is an eruption.

The "lid" is a column of hot water standing in the vent, weighing down the even hotter water beneath. A slight lessening of that weight – a splash or a brief surge – is all it takes to trigger an eruption.

Sunset illuminates an eruption of Clepsydra Geyser (above).

Boiling "mudpots" (opposite, top) and the stench of sulfur are common in places where groundwater meets hot rocks at shallow depths.

Old Faithful (opposite, bottom) now erupts approximately every 78 minutes. Recent earthquakes altered the geyser's hydraulics, lengthening the interval.

Yellowstone geysers fall into two categories. **Old Faithful** is a cone type, meaning that it emerges as if from a nozzle. **Beehive**, **Lion**, **Castle**, **Riverside**, and **Daisy**, other examples, have continuous streaming jets of water in the early stages of eruption, gradually giving way to bellowing steam. Fountain geysers – including **Grand**, **Giantess**, **Great Fountain**, **Clepsydra**, and **Echinus** – emerge from a hot pool, usually in explosive bursts that eventually empty the pool. In both cases, eruptions begin when the critical balance between the weight of overlying water and the growing pressure from beneath tips in favor of the pressure.

The biggest Yellowstone geysers are found in the Upper, Middle, and Lower Geyser Basins along the Firehole River and in Norris Geyser Basin at the head of the Gibbon River. Norris claims the park's hottest ground and the world's biggest active geyser, **Steamboat**. Capable of hurling water 300 feet high, its eruptions are rare and unpredictable events. Be patient – you may have to wait for months.

the intervening mountains.

When young, the caldera must have been a dramatic sight, a huge oval depression ringed by steep walls, its still-hot floor belching vapor. However, lava flows filled in large portions of the crater, obscuring its walls and spilling out of the park toward the southwest. Glaciation further bulldozed the surface. Yellowstone Lake fills part of it, but the lake's southern arms reach miles beyond the former rim. Only a geologic map shows the caldera clearly. The edge is defined by volcanic tuff, a porous consolidated volcanic ash, while the interior is largely rhyolite deposited by the lava itself as a heavy, granite-like rock.

Don't leave Mount Washburn without enjoying the **Absarokas**. Their snowy summits, marking Yellowstone's eastern boundary, rise beyond the rim of the caldera on the far side of a huge forested expanse. This is the **Mirror Plateau**, one of the resurgent domes that portend future volcanic activity.

From here you can also see the rim of the **Grand Canyon of the Yellowstone** as a dark gash in the pine-clad surface. Up close, the canyon is unexpectedly beautiful, not dark at all, thanks to geothermal altering of the rhyolite through which it was carved. Heat and associated chemicals transformed the gray rock into yellows, browns, and reds. They also softened the rock, letting it erode

that was saturated with superheated water. No longer held under pressure, the groundwater flashed into steam like a bomb.

Another explosion crater, **Pocket Basin**, is perhaps the most distinct in the park. It lies just off the **Fountain Flat Drive** in the **Lower Geyser Basin**. If you park at **Ojo Caliente Hot Spring** and walk up the low hill across the road, you'll stand on the rim of the crater among fragments of rock shattered by the explosion.

more easily than the unaltered rhyolite upstream. The canyon's two great waterfalls mark the junction of hard and soft rock.

Glaciers and hot springs had other interactions in the Lower Geyser Basin. West of

Fire and Ice

To see a more obvious caldera, follow the road from **West Thumb Geyser Basin** toward Old Faithful about one mile to a panoramic view of Yellowstone Lake and the distant Absarokas. **West Thumb**, the closest bay of the lake, is a caldera formed less than 200,000 years ago by a secondary eruption within the Lava Creek caldera. Filled with water, it retains its rounded shape and is about the same size as another regional caldera, Oregon's Crater Lake.

Now look down. That pond tucked into the forest between the viewpoint and West Thumb is one of Yellowstone's oddest oddities. Called **Duck Lake**, it is a thermal explosion crater, one of several in the park. These occurred during late glacial times when areas of hot ground melted the ice directly above, creating deep lakes locked within the glacier. If something breached the confining ice, the lakes would drain abruptly, releasing enormous weight from ground

Fountain Paint Pot stand two rounded hills, **Twin Buttes**. These are thermal kames, created when hot springs melted holes in a glacier, and surface streams carrying rocks and sand poured into the holes.

Yellowstone's premier attraction, **Old Faithful Geyser**, is a fine sight early in the morning when the steam is dense. But there are many other geysers of all sizes, flanked by thousands of hot springs and steam vents. Thermal areas dot the entire park, with the densest concentrations along the **Firehole** and **Gibbon Rivers**. The most important are near the road, but many others exist in backcountry areas.

Geysers seem magical and mysterious, but with a good heat source, plenty of water, and lots of pipe, you could build a working model. In Yellowstone's case, the heat comes from the ground, the water from the sky, and the pipe … well, the pipe is more complicated. The plumbing is a combination of fractured bedrock into which groundwater can seep; glacial sand and gravel, which act as a sponge to hold yet more water; and a cap of thermally deposited silica dioxide called sinter, which is a white rock that seals the basin and allows pressure to escape through a limited number of openings. Without their sinter caps, geyser basins would be little more than steaming gravel beds.

Mineral deposits color the clear water of Morning Glory Pool (above).

Mineral-laden water builds unusual deposits (right) as it bubbles out of the ground in Midway Geyser Basin and streams into the Firehole River.

The Tetons (right and bottom) are fault-block mountains. As one side of the fault rose and tilted, the other subsided. Glaciers later sharpened the peaks and widened the valleys between them.

Terraces of Travertine

Northern Yellowstone escaped the total remodeling done to most of the park. Its landscape, through dramatic, was built by events typical of the Rockies – ancient seabed deposition, then regional uplift, followed by periods of volcanism and erosion. The result is a more conventional mountain landscape. The Absarokas are old volcanoes, but the **Gallatins** are Mesozoic and Paleozoic sediments on a Precambrian base, meaning that they were laid down between 65 and 570 million years ago on ancient bedrock. Their geologic inventory includes fossils, landslides, glacial deposits, petrified wood, abandoned streambeds, and drained Ice-Age lakes.

   Mammoth Hot Springs is the exception to this ordinary mountain landscape. The steaming white hillside covered with steplike terraces is solid travertine – calcite deposited by hot springs. The mineral's

The Youthful Tetons

Once seen, the **Teton Range** is not forgotten. The flat valley called Jackson Hole, the absence of foothills, and the clean sweep of gray rock soaring above a chain of glittering lakes set it apart from other ranges. Its highest point is **Grand Teton**, at 13,770 feet. Flanking it are six other peaks over 12,000 feet and many subsidiary crags, making this one of North America's most dramatic geological statements.

The mountains are still rising along a fault between two blocks, one rising as the other sinks. Together they account for 30,000 vertical feet of displacement. We don't see all of that because erosion has removed much of the upraised sedimentary layers while filling the valley below with washed-off debris.

Some uplifted sediments remain. Limestone and sandstone still cover the western slope of the range. And on the summit of **Mount Moran**, you can see a small patch of sandstone corresponding to a layer more than four miles beneath the valley. The dominant material of the mountains, however, is hard old basement rock – schist and gneiss wrapped around granitic intrusions. It's good rock for climbing and great rock for scenery.

We owe the Tetons' craggy appearance to glaciers, which sharpened their summits over the last two million years. During the most recent glaciation, the Yellowstone ice cap pushed into Jackson Hole but did not cover the Tetons. The mountains grew their own smaller glaciers. These poured down the canyons and gouged moraine-rimmed basins along the foot of the range. Today, these basins contain a string of lakes, including exquisite **Jenny Lake**.

The Tetons are young, as shown by several datable geological events. The most compelling is an ash-flow deposit, five million years old, found in Jackson Hole but originating in Idaho. It could never have flowed up and over the mountains – which strongly implies that they weren't here at the time.

Such quick mountain-building is no gentle process. Although the Teton fault hasn't produced an earthquake in historical times (less than 150 years), geologists warn that the uplift is ongoing – and that when quakes do occur here, they tend to be major.

source, **Terrace Mountain**, rises above and behind the hot springs. Made of limestone, the mountain is gradually being dissolved and redeposited by the action of hot water.

East of Mammoth Hot Springs, the road parallels the rugged **Black Canyon of the Yellowstone**. Rocks vary from Absaroka volcanics to glacial deposits to welded tuff thrown out by the caldera eruptions. At **Tower Junction**, continuing eastward, you cross the Yellowstone River and pick up the **Lamar River** where it flows through a narrow canyon. There's an interesting glacial story here. **Lamar Valley** is the bed of an Ice-Age lake periodically dammed by advancing glaciers. It would fill with water until the ice could no longer hold it back. Then the ice dam would release and send a huge flood racing downstream through **Black Canyon** and out past **Gardiner**.

Older calamities are recorded off **Specimen Ridge** on the south side of Lamar Valley. Here, abundant petrified wood reveals how eruptions during the Absaroka volcanic period repeatedly buried existing forests, making a layer cake of buried woodlands. Look for a petrified tree one mile

west of Tower Junction at the end of a short side road. An iron fence protects the ancient stump from souvenir hunters.

Another site worth visiting is outside the park. The **Beartooth Mountains**, consisting of Precambrian basement rocks 2.7 billion years old, stand in contrast to the softer, younger volcanics. The Beartooths tilted as they rose, causing overlying sediments to slide off toward the southeast. You can see this well along the **Chief Joseph Scenic Byway** (Route 296), and there's a particularly fine view of both the Absarokas and the Beartooths from **Dead Indian Pass**. It would be hard to find a region whose geologic history is laid out so neatly.

Columnar joints (above) formed in the Pleistocene basalts that compose the Sheepeater Cliffs south of Mammoth Hot Springs.

Iron Spring (left), colored by oxidized iron in the water, lies along the Gibbon River between Norris and Madison.

Lower Yellowstone Falls (opposite) plunges 308 feet into a rhyolite canyon that was softened and colored by hot groundwater.

TRAVEL TIPS

DETAILS

When to Go

The park is most inviting, and most crowded, in summer. Weather in July and August is warm and pleasant, with highs around 90°F and cool nights. Winter is harsh, with extremes well below 0°F. Most park roads are closed by snow from November to April.

How to Get There

Commercial airlines serve Cody and Jackson, Wyoming; Bozeman and Billings, Montana; and Idaho Falls, Idaho. The West Yellowstone, Montana, airport is open June to early September.

Getting Around

A car is the most convenient way to get around; rentals are available at the airports. Bus tours are offered by AmFac Parks and Resorts, 307-344-7311, and National Park Tours–Grayline, 307-733-4325.

Handicapped Access

Visitor centers, some campsites, and many park-run activities are accessible. Detailed information is available from the park's handicapped-access coordinator; write to Box 168, Yellowstone National Park, WY 82190.

INFORMATION

Jackson Hole Chamber of Commerce

532 North Cache Street, P.O. Box 550, Jackson, WY 83001; tel: 307-733-3316.

Yellowstone National Park

P.O. Box 168, Yellowstone National Park, WY 82190-0168; tel: 307-344-7381.

CAMPING

There are 12 campgrounds in the park; seven are open in summer and fall. Only one, Mammoth, is open year-round. Reservations can be made for five sites by calling AmFac Parks and Resorts, 307-344-7311; the rest operate on a first-come, first-served basis. A backcountry permit is required for all overnight backpacking trips and can be obtained in person at most ranger stations and visitor centers no more than 48 hours in advance. For a campsite reservation form, contact the Backcountry Office, 307-344-2160.

LODGING

PRICE GUIDE – double occupancy

$ = up to $49 $$ = $50–$99
$$$ = $100–$149 $$$$ = $150+

Canyon Lodge

P.O. Box 165, Yellowstone National Park, WY 82910; tel: 307-344-7311.

This 79-room, wood-frame lodge is flanked by an additional 540 cabins, each containing four or more units with private bathrooms. The complex is on the Loop Road about half a mile from the Grand Canyon of the Yellowstone River. The main building contains a dining room, cafeteria, snack shop, and lounge. Open early June to early September. $$–$$$

Lake Yellowstone Hotel and Cabins

P.O. Box 165, Yellowstone National Park, WY 82910; tel: 307-344-7311.

On Lake Yellowstone in the southeast part of the Loop Road, this four-story hotel, the park's oldest building, was erected in 1891 and recently renovated. The main building has more than 190 rooms with lake or mountain views. One hundred cabins provide basic lodging, each with two double beds, some with private baths. A dining room, deli,

marina, and tours are available. Open mid-May to early October. $$–$$$$

Old Faithful Inn

P.O. Box 165, Yellowstone National Park, WY 82190; tel: 307-344-7311.

A National Historic Landmark set near the famous geyser, this 325-room log hotel was built in 1904. It has a massive, four-sided fireplace in the lobby and offers simple but comfortable accommodations, a dining room, and lounge. Open early May through mid-October. $$–$$$$

Roosevelt Lodge and Cabins

P.O. Box 165, Yellowstone National Park, WY 82190; tel: 307-344-7311.

At Tower Junction in the northeast section of the Loop Road, Roosevelt Lodge offers rustic accommodations in "Rough Rider" style, commemorating Theodore Roosevelt, who owned a cabin nearby. The 62 cabins offer basic comfort, with two double beds and a wood-burning stove. Sink, toilet, and shower facilities are in communal bathrooms; there is no running water in the cabins. Open mid-June to early September. $–$$

TOURS

Dozens of guides and outfitters work in the Yellowstone region. A list of those licensed to operate in the park is available from the Yellowstone Visitor Service, 307-344-2107. Here are just a few that specialize in natural history:

Sierra Safaris Wilderness Tours

P.O. Box 963, Livingston, MT 59047; tel: 406-222-8557 or 800-723-2747.

Guides lead custom-designed explorations of Yellowstone National Park and surrounding areas, including an adventure safari that focuses on geothermal features and wildlife. Hikes are tailored to the guests' interests and abilities.

Timberline Adventures

7975 East Harvard Avenue, Suite J, Denver, CO 80231; tel: 303-368-4418 or 800-417-2453.

A six-day hike through Yellowstone and Grand Teton explores geothermal features, waterfalls, and canyons. The inn-to-inn hike features overnight stays at Old Faithful Inn and Jackson Lake Lodge.

The World Outside

2840 Wilderness Place, Boulder, CO 80301; tel: 303-413-0938 or 800-488-8483.

Yellowstone's canyons and waterfalls, hidden hot springs, steaming fumaroles, geysers and bubbling mud-pots are featured in a six-day hike through the park.

MUSEUMS

Museum of the Rockies

600 West Kagy Boulevard, Bozeman, MT 59717; tel: 406-994-3466.

The museum presents a detailed picture of the Yellowstone region's ecosystem, past and present, including an outstanding dinosaur fossil collection.

Norris Geyser Basin Museum

P.O. Box 168, Yellowstone National Park, WY 82190; tel: 307-344-2812.

The Norris Geyser Basin is the hottest and most volatile geothermal area in the park. Exhibits explain how geysers are created and how they operate. Two loop trails provide a safe route for viewing the geysers and hot springs of the Porcelain and Back Basins.

Excursions

Florissant Fossil Beds National Monument

P.O. Box 185, Florissant, CO 80816; tel: 719-748-3253.

The park preserves a deposit of 33- to 35-million-year-old fossil plants and insects about 20 miles northwest of Colorado Springs. Two trails, Walk Through Time and Petrified Forest Loop, lead to fossil-bearing shale beds and the remains of a "petrified forest."

Fossil Butte National Monument

P.O. Box 592, Kemmerer, WY 83101; tel: 307-877-4455.

A 50-million-year-old lakebed in what is now southwest Wyoming contains one of the richest deposits of fossils in the world. Fish are the most abundant, but the bed also preserves plants, insects, and other animals over a period of two million years.

Glacier National Park

P.O. Box 128, West Glacier, MT 59936; tel: 406-888-7800.

Push came to shove in the northern Rockies starting about 170 million years ago. That's when a slab of billion-year-old Precambrian sedimentary rocks several thousand feet thick was rammed more than 50 miles eastward over much younger Cretaceous rocks by a thrust fault. For gorgeous vistas, take either the Going-to-the-Sun Road or U.S. Highway 2.

Rocky Mountain National Park

Park Headquarters, Estes Park, CO 80517; tel: 970-586-1206.

Glacial sculpturing is the main feature of the park, whose rocks are largely Precambrian granites, gneisses, and schists. In a few places, notably along Trail Ridge Road, you can see mid- to late-Tertiary volcanic ashflows. In the high elevations, a few living glaciers continue the ice's eternal work.

Zion
National Park
Utah

CHAPTER **11**

Fifteen hundred feet above **Zion Canyon**, a red-tailed hawk calls to her mate from a ledge in a side canyon, her eerie *skree* echoing off pale sandstone. As far as the eye can see, an electric-blue sky stretches over a frozen ocean of twisted stone knobs, buttes, and soaring temples deeply fissured into crags and finger rocks. Spring winds eddy over sand pockets frilled with purple milkvetch and stir the branches of high-country ponderosa pines. The trees creak in their fragile rock moorings as they continue their ancient conversation with the elements. ◆ In a thousand nooks and crannies throughout Zion, water moving over and through sandstone dislodges pebbles and rocks. Skittering down the cliffs into the streambeds, they add to the heavy payload of the **Virgin River** and its tributaries, the master sculptors of Zion. ◆ If you've never thought much about rocks before, you will after a visit to **Zion National Park**. This 229-square-mile park in southwestern Utah offers geology so glorious that you ache

Soaring cliffs, set ablaze by the rising sun, are one of many attractions that lure amateur geologists to this spellbinding slickrock wilderness.

to commune with it, to feel its smooth slickrock under your boots, to watch shadows and light play across cliff faces. You leave with a whole new vocabulary and appreciation for the beauty and fickle ways of stone. ◆ For hikers slowly ascending the narrow ridge of **Angels Landing** on the **West Rim** of Zion Canyon, the experience is certainly beyond comparison – if only because an attack of vertigo or a foot in the wrong place could be fatal. After negotiating the steep trail from the Grotto Picnic area and the numerous sharp switchbacks of Walters Wiggles, many visitors are happy to catch their breath and enjoy the view at Scout Lookout.

Angels Landing stands high above the Virgin River, which is carving downward through the sandstones, shales, and limestones of Zion Canyon.

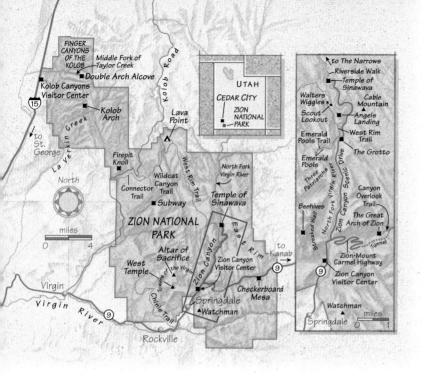

Through all flows the Virgin River. Its waters have the consistency of pea soup and smell of rotting winter vegetation as they gurgle over and around boulders, slowing to drop sediment on bends. When raging in flood they carve even deeper into the layer cake of sedimentary rocks laid down in 250 million years of inland oceans, deserts, rivers, braided streams, and mudflats.

other formations in the park, by an enthusiastic clergyman). At the park's western entrance, the great hulk of the **Watchman** broods above the visitor and transit center.

Canyon Overview

Zion can be seen from many angles and at many speeds, but avoid the temptation to rush. Get up high, then down low, hide in canyons and stand atop ridges. Get a feel for the park's 4,000-foot deserts, 8,000-foot high country, verdant riparian corridors, and many microclimates.

For a geological overview – one not requiring strenuous hiking and a head for heights – enter Zion via the East Entrance from Kanab and hike the one-mile round-trip **Canyon Overlook Trail**. It passes alcoves and seeps and **Pine Creek Narrows** to end up atop the **Great Arch of Zion**, a huge blind arch above cottonwood-lined Pine Creek. From here, you have an ace view of **Zion Valley**, the switchbacks looping down into the valley, and the soaring **West** and **East Temples**, the **Towers of the Virgin**, the **Streaked Wall**, the **Beehives**, and other prominent features of lower Zion Canyon.

To the left is the 1.1-mile **Zion-Mount Carmel Tunnel**, a masterpiece of engineering completed in 1930. Old-timers will tell you

Others with a taste for high adventure press on up the promontory, steadying themselves on chain rails and refusing to look down.

Across Zion Canyon on the **East Rim** is **Cable Mountain**, where a century ago young Mormon settler David Flanagan secured a cable to send timber from rim to valley floor. Off to the left, out of sight, is the **Temple of Sinawava**, a 2,000-foot sandstone monolith, and the **Riverside Walk Trail**. It leads to the long narrow slot canyon carved by the **North Fork of the Virgin River**. To the right, down canyon, are the conferring heads of the **Three Patriarchs** (named, along with

Zion Canyon Stratigraphy

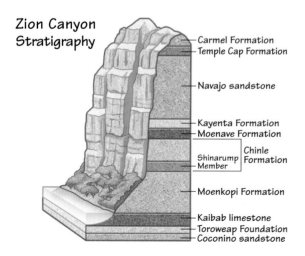

- Carmel Formation
- Temple Cap Formation
- Navajo sandstone
- Kayenta Formation
- Moenave Formation
- Shinarump Member — Chinle Formation
- Moenkopi Formation
- Kaibab limestone
- Toroweap Foundation
- Coconino sandstone

which forces groundwater to emerge where it touches the overlying porous sandstone. The water destabilizes the sandstone, which falls away in vertical sheets like the layers of an onion. Elsewhere soft rocks erode into hoodoo columns.

Zion's caprocks are fossil-bearing Carmel limestone and the ruddy Temple Cap Formation, both laid down in an ancient sea about 175 to 180 million years ago. Below these is the park's signature rock – the massive, creamy Navajo sandstone. It sits above the 200-foot-thick, maroon Kayenta shale, which occasionally yields dinosaur tracks. Below that lies the red marine Moenave siltstone.

The Moenave sits atop the multihued and fossil-rich Chinle Formation, a crumbling mix of mudstone, siltstone, and volcanic ash. This is the rock found at Petrified Forest National Park in Arizona; it also contains

that in days gone by visitors used to be able to stop their cars at the "windows" overlooking the canyon. Those days are long gone now, with more than 2.5 million people visiting Zion annually and increased traffic on Highway 9. But the tunnel still offers brief, teasing glimpses of the canyon before you exit to daylight on the other side.

Geologic Layers

The park's main geological formations can be seen from the **Canyon Overlook Trail**. Looking closely at high landmarks like the West Temple, you can begin to read the rocks. Thick, hard limestone and sandstone beds form sheer cliffs, while thin, compacted shales, siltstones, mudstones, and other freshwater sediments are softer and erode into slopes. The shales are relatively impermeable,

Checkerboard Mesa (above) is a large exposure of Navajo sandstone, weathered to reveal intricate crossbedding, a legacy of its birth as a vast field of sand dunes.

A stream cuts into sandstone (right), enlarging a natural joint in the rock.

Zion Narrows (right) is a slot canyon where the Virgin River has sliced down through the rock, making a shadowed passage only as wide as the river.

Snowmelt sends a waterfall gushing over the wall at Zion Narrows Gateway (below).

the great uranium deposits that helped fuel the United States' nuclear weapons program. Petrified wood may be viewed in the Chinle along the 14-mile round-trip **Chinle Trail** between Rockville and Springdale. The Chinle's bottom member is the 100-foot-thick Shinarump Conglomerate, a coarse rock made up of sand and gravel and prominently light tan in color. Finally, at river level, is the Moenkopi Formation, a colorful rock up to 1,800 feet thick. It was laid down as red, brown, and pink bands of sand-stone, shale, limestone, and gypsum from streams as well as oceanic and coastal plains.

Ancient Dunes

The soaringly beautiful Navajo sandstone is the rock that gives Zion its trademark uniform mesas and vertiginous cliffs, red-to-rose-to-buff coloration, and distinctive alcoves, arches, joints, clefts, and fingers. The sandstone originated some 175 million years ago, during a time of great aridity, when a 3,000-foot-deep sand-dune desert

Discrete layers of sandstone
(right) record individual storms or
other depositional events.

Ancient petroglyphs (below) were
created by pecking away the dark,
weathered surface of the rock.

lay across the West from
Wyoming to Nevada, some
150,000 square miles in all.
Today, the Navajo sandstone
reaches its greatest thickness in Zion – 2,400 feet at
the Temple of Sinawava.

The sand was borne on
winds that chopped and
changed direction, leaving
behind cross-bedded dunes
too mobile to support life
except in the interdunal
areas, where dinosaurs
occasionally roamed. As
the climate changed,
encroaching seas covered
the dunes and marine sed-
iments cemented the sand
grains with limey calcite
and iron-rich hematite.

The most dramatic
"petrified dunes" lie at
Checkerboard Mesa, an eolian monolith
with a distinctive cross-hatched surface
that gave the mesa its name. Deep, sandy
washes nestle at the petrified dunes' feet,
evidence of the relentless power of erosion.
Set free again after being locked up for
ages in the sandstone, the grains of sand
are picked up on prevailing winds and
redeposited as modern-day dunes at such
places as **Coral Pink Sand Dunes State
Park**, southeast of Zion.

For unknown reasons, the Navajo takes
on different hues in different areas. One
theory suggests that groundwater moving
through the porous rock washes out the
minerals, leaving crumbling pale caprocks
above dense red bases.

How then to explain the rocks in Zion's

Kolob Unit, far to the west? Here the cliffs are
uniformly red and seem to catch fire at sunset (it's the best photo opportunity in Zion).
To reach this part of the park by car, drive
west to Interstate 15, then head north and
turn at the Finger Canyons of the Kolob exit.

Slot Canyons

These are the Southwest's secret places – the twisting water-carved slot canyons that honeycomb northern Arizona and southern Utah like catacombs beneath a giant cathedral.

Far below the canyon rim, a soft carpet of sand lies underfoot, muffling footsteps. In some places, rockfalls slow travel. The slightest whisper is magnified here, echoing off vertical sandstone walls that may reach thousands of feet high and stand less than a foot apart. Glimpses of sky appear, then disappear. Light is a reflected memory, illuminating taffy swirls of sandstone. It is a world, writes photographer Michael Fatali, who specializes in slot canyons, that "suggests qualities and objects that in reality do not exist. Only our imagination guides us."

Slot canyons, among the most remarkable and beautiful geological features of canyon country, exist in great numbers where streams have cut straight down through sandstone like a knife through butter. Unlike ordinary erosion, in which rocks of differing compositions and hardness weather into stepped-back cliffs and hoodoos, slot canyons form when the bedrock is uniform from top to bottom. Water entering joints finds little resistance, and weathering accelerates during times of runoff. Then, sediment- and debris-laden flash floods, fueled by unseen storms upcountry, roar through the narrow recesses at the speed of a freight train with little or no warning.

The most celebrated slot canyons – **Zion Narrows** in Zion National Park, **Antelope Canyon** on the Navajo Reservation near Page, Arizona, and nearby **Paria Canyon** – see a fair number of visitors. Others, such as the many slots in **Grand Staircase-Escalante National Monument** and **Glen Canyon National Recreation Area**, offer rock scramblers and canyon lovers plenty of adventure and solitude.

But take care. Flooded canyons can snare a human life as quickly as a Venus flytrap. Be wary during the summer monsoon season, and try to plan a visit in early summer or fall when there's less danger of flash floods.

A hiker (left) explores an Arizona slot canyon.

Zion's monoliths, such as the Towers of the Virgin (right), stem from its thick layers of relatively homogeneous rock.

Water (opposite, bottom) trickling through Navajo sandstone is forced out when it reaches the less-permeable Kayenta shales and siltstones of Weeping Rock.

into the backcountry via **La Verkin Creek** takes you to 310-foot **Kolob Arch**, considered the world's longest arch.

Trail Options

Long hikes into the Kolob drainage require planning, permits, and warm weather. This is particularly true if you plan to hike out of La Verkin Creek onto the 10,000-foot **Kolob Terrace** along the **West Rim Trail** and down into Zion Canyon. A scenic backway turning off Highway 9 at Virgin, west of Zion, leads to the Zion high country on the 11,000-foot Markagunt Plateau (it is closed by snow at Lava Point, November to May). Recent volcanic activity can be seen on the Kolob Terrace at places like **Firepit Knoll** and **Lava Point**.

Plan any trip here with a backcountry ranger at the **Zion Canyon Visitor Center**. All overnight trips and some day hikes require permits. However, the 5.8-mile **Wildcat Canyon Trail** and the 4-mile **Connector Trail**

The country here is wilder and less visited, a maze of dramatic box canyons and deep creek beds. The **Middle Fork of Taylor Creek**, a 5.4-mile round-trip, offers easy hiking to the unusual **Double Arch Alcove**, while a popular overnight hike seven miles

raging floodwaters, and bubbling springs shaping the land in front of you. Water also washes minerals out of the rocks, causing flaring red streaks on landmarks like the **Altar of Sacrifice** and white tufa around seeps. Particularly noticeable is the dark, shiny desert varnish on exposed cliff faces. Oxidized minerals like iron and manganese become fixed on rock surfaces by dust and bacteria and form this dark patina, common throughout the Southwest.

offer moderate nonpermit day hikes for anyone wanting dramatic high-country views and solitude.

The Kolob Terrace offers a view of Zion behind the scenes and generally not for anyone without a good degree of fitness and route-finding skills. This holds especially for the extremely strenuous, six-mile **Left Fork of North Creek Trail**. It leads to the **Subway**, a passage created by water seeping from the contact between the Navajo sandstone and the Kayenta shale. This type of cool alcove occurs elsewhere in the park, including along the **Canyon Overlook Trail** and at the shady **Emerald Pools Trail** in Zion Canyon. Such seeps and springs support wildlife and hanging gardens of maidenhair fern and monkeyflower, and offer respite during the searingly hot summers that assail Zion at its lower elevations.

Visit Zion during spring runoff or summer monsoon season and you'll witness waterfalls,

TRAVEL TIPS

DETAILS

When to Go

Spring and fall are the most pleasant, though you may want to avoid the crowds during spring break. Summer temperatures exceed 100°F in lower elevations; thunderstorms and flash floods occur from July to September. Snow closes the high country from late September to late May, and low-elevation trails may be icy and slippery.

How to Get There

Major airlines serve Salt Lake City, about six hours north of Zion National Park via Interstate 15. Commuter airlines serve nearby St. George, an hour from the park via Highway 9.

Getting Around

A car is essential. Car rentals are available at the airports. High-clearance four-wheel-drive vehicles are often necessary on unpaved country roads.

Backcountry Travel

Overnight trips and some day trips into the park's backcountry require permits. Quotas and fees apply; call the park for details.

Handicapped Access

The visitor centers, shuttle bus, and some campsites and paved trails are wheelchair-accessible. Contact the park's handicapped access coordinator for details.

INFORMATION

Kane County Travel Council

78 South 100 East, Kanab, UT 84741; tel: 435-644-5033.

St. George Area Chamber of Commerce

Old Washington County Courthouse, 97 East St. George Boulevard, St. George, UT 84770; tel: 435-628-1658.

Zion National Park

Springdale, UT 84767; tel: 435-772-3256 (Zion Canyon Visitor Center) or 435-586-9548 (Kolob Canyons Visitor Center).

CAMPING

Camping is available in South Campground and Watchman Campground just inside the southern entrance to the park. Arrive by early morning; the campgrounds fill fast. Campsites at Watchman Campground may be reserved at least 24 hours in advance; call 800-365-2267. Six primitive campsites are available from May to October at Lava Point on the Kolob Terrace.

LODGING

PRICE GUIDE – double occupancy

$ = up to $49 $$ = $50–$99

$$$ = $100–$149 $$$$ = $150+

Canyon Ranch Motel

668 Zion Park Boulevard, Springdale, UT 84767; tel: 435-772-3357.

This budget motel has two- and four-unit cottages with great views of Zion. Rooms have one or two beds; some have kitchenettes. A swimming pool and Jacuzzi are on the premises. $–$$

Cliffrose Lodge and Gardens

281 Zion Park Boulevard, P.O. Box 510, Springdale, UT 84767; tel: 800-243-8824 or 435-772-3234.

The lodge is set on the Virgin River near the park's south entrance and offers 30 rooms and six suites. $$–$$$

Desert Pearl Inn

707 Zion Park Boulevard, P.O. Box 407, Springdale, UT 84767; tel: 888-828-0898 or 435-772-8888.

The inn features Pueblo-style architecture and spacious rooms with microwaves, refrigerators, wet bars, pull-out sofa beds, and views of the park and Virgin River. A large pool is on the premises. $$–$$$

Zion Lodge

AmFac Parks and Resorts, 14001 East Illiff Avenue, Suite 600, Aurora, CO 80014; tel: 303-297-2757 or 435-772-3213 (in the park).

The 121-room lodge, nestled at the base of towering cliffs near the Virgin River, was rebuilt after a fire in the 1960s. The building now has motel-style rooms. The renovated cabins were built in the 1920s and have private porches, stone fireplaces, log beams, and two double beds. A restaurant, gift shop, and snack bar are on the premises. The lodge is open year-round. $$

Zion Ponderosa Ranch Resort

P.O. Box 5547, Mount Carmel, UT 84755; tel: 435-648-2700 or 800-293-5444.

Situated on Zion's quiet east rim, this new resort offers log cabins, campsites, and excellent food. Activities include hiking, horseback riding, wagon rides, climbing, guided tours, and a daily children's camp in summer. $$–$$$

TOURS

Bike Zion

1458 Zion Park Boulevard, Springdale, UT 84767; tel: 800-475-4276 or 435-772-3929.

Guided half-day and full-day hikes explore Zion and nearby areas. Multiday backpack trips are also available, including canyoneering in slot canyons with a rappelling instructor.

Canyon Trail Guides

P.O. Box 128, Tropic,
UT 84776; tel:
435-679-8665.

This outfit leads
guided horseback
rides through Zion
from March to
October. Reservations
are advised.

Zion Adventure Company

36 Lion Boulevard,
P.O Box 523,
Springdale, UT
84767; tel:
435-772-1001.

The company provides gear and
information for canyon excursions
and backcountry hikes. Shuttle
service in Zion and guided
canyoneering and adventure
day camps for children are also
available.

MUSEUMS

The Grand Circle: A National Park Odyssey

Obert C. Tanner Amphitheater,
225 South 700 East, St. George,
UT 84770; tel: 435-652-7994.

This one-hour multimedia presen-
tation offered by St. George's
Dixie College covers the geology
and spectacular parks of Utah
and northern Arizona.

Zion Canyon Cinemax Theatre

145 Zion Park
Boulevard,
P.O. Box 206,
Springdale, UT
84767; tel:
435-772-2400.

Films on the nat-
ural and cultural
history of Zion are
shown hourly on
the theater's giant
screen.

Excursions

Bryce Canyon National Park

*P.O. Box 170001, Bryce Canyon, UT
84717; tel: 435-834-5322.*

Bryce Canyon is a testament to the power
of erosion. Over millions of years, water
has sculpted the bizarre hoodoos, spires,
and balanced rocks from a bed of soft
Claron Formation limestone, which is tinted
a brilliant orange and yellow by iron and
manganese. Not one canyon but 14
separate amphitheaters, the park may be
viewed from a 13-mile scenic drive and
nine different trails.

Cedar Breaks National Monument

*2390 West Highway 56, Suite 11,
Cedar City, UT 84720; tel:
435-586-9451.*

This highly eroded amphitheater
on the western edge of the 10,000-
foot Markagunt Plateau was
called Circle of Painted Cliffs by the
southern Paiute but renamed by
Mormon pioneers for its lack of
trees. Steeper and more colorful
than Bryce Canyon, and still pure
wilderness, Cedar Breaks is a hidden,
high-country gem surrounded by
volcanic lava flows and alpine
meadows that explode with wildflowers in summer.

Snow Canyon State Park

*P.O. Box 140, Santa Clara,
UT 84765-0140; tel: 435-628-2255.*

Snow Canyon, 11 miles northwest of St.
George, is one of Utah's most geologically
interesting state parks and was used as
a backdrop in several John Wayne
Westerns. It's surrounded by sheer, eroded
Navajo sandstone cliffs and set off
dramatically by black basaltic lava flows
associated with the volcanic Pine Valley
Mountains to the east. Enjoy the park on
horseback or by hiking to lava caves,
arches, and petroglyphs. The campground
is nestled amid slickrock.

Canyonlands National Park
Utah

CHAPTER 12

W hen you really get smitten by rocks –
when you can't look at a cliff without thinking of the name of the formation,
its age, its thickness – there's only one solution. Head immediately to
Canyonlands National Park in southeastern Utah and give yourself over
to the obsession until it runs its course.   ◆   You'll sit on warm rocks,
staring at 100-mile horizons until the dark shapes of peaks, mesas, and
anthropomorphic formations dance on your retinas. Sun and clouds move
across a bolt-blue sky, playing shadows on rocks a thousand feet below.
Eventually you'll be drawn into the desolate beauty of the canyons, where
the geometric orderliness of the rim country quickly proves an illusion. Down
there it's a topsy-turvy world of arches, bridges, reefs, spires, and balanced
rocks.   ◆   This 527-square-mile desert **A labyrinth of canyons,**
park is huge and wild. It has few paved **arches, and dreaming**
roads and just a scattering of abandoned mining **spires sprawls across**
and ranch trails. Even the cairns marking foot **Utah's red-rock country.**
trails require attention to avoid a wrong turn. Plan any trip here carefully –
especially in summer, when temperatures soar above 100°F in the lower
elevations. In winter, ice, snow, and frigid temperatures are equally threaten-
ing to those unaccustomed to desert extremes.   ◆   The park is divided
into three districts – **Island in the Sky**, the **Needles**, and the **Maze**. The
Colorado and **Green Rivers**, which flow in deep canyons on the east and
west sides of Island in the Sky, form an unofficial fourth river district. For an
overview, make for **Grand View Point** on 6,000-foot-high Island in the Sky,
off Highway 191 west of Moab. The role of water and time in creating this
geological miracle becomes more understandable with the advantage of height.

Washer Woman Spire, catching the
sunset near Island in the Sky, is composed
mostly of massive Wingate sandstone,
eroded along the rock's joints.

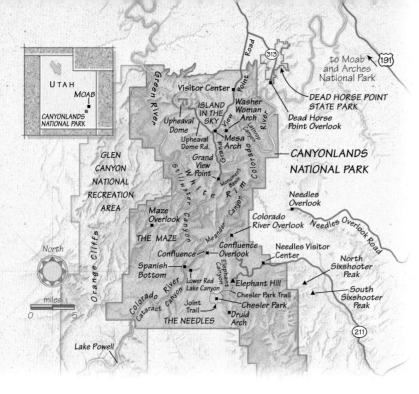

View from the Top

Canyonlands lies within the Paradox Basin and is encircled by mountains whose lava cores never broke the surface: clockwise from the northeast, the **La Sals** at two o'clock, the **Abajos** at six o'clock, the **Henrys** at eight o'clock, the **Book Cliffs** behind, at noon. To the south lie the banded, dreaming spires of the Needles, east of the Colorado River. They are mirrored on the west side of the river by the even more convoluted Maze, canyons so remote and labyrinthine even the ravens are

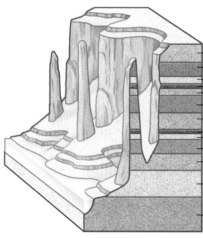

Canyonlands Stratigraphy

- Morrison Formation
- San Rafael Group
- Navajo sandstone
- Kayenta Formation
- Wingate sandstone
- Chinle Formation
- Moenkopi Formation
- White Rim sandstone
- Organ Rock Formation
- Cedar Mesa sandstone
- Elephant Canyon and Halgaito Formations
- Honaker Trail Formation
- Paradox Formation

said to ask for directions.

Island in the Sky and the neighboring cliffs are composed of fractured pink Wingate sandstone topped by hard red Kayenta shale caprock and, in places, Navajo sandstone. They form lonely palisades atop the crumbling Chinle and Moenkopi slopes rich in uranium – deposits that helped nearby Moab boom in the 1950s but which left a legacy of cleanup problems outside the town. Midway between the Island and the rivers is a huge terrace composed of hard, thin, marine-deposited White Rim sandstone.

The White Rim sandstone slows the erosion of the underlying soft, chocolate-hued Organ Rock shale and banded Cedar Mesa sandstone, and the Elephant Canyon Formation. It offers 100 miles of adventurous four-wheel driving and mountain biking on the **White Rim Trail**, a big draw in spring. The White Rim also provides the main access to the rivers via inner canyons like Lathrop, named for an early rancher, and Monument Basin, filled with totems, balanced rocks, arches, and house-sized boulders tipped by gravity.

Shaped by Water, Underlaid by Salt

When it gets gusty up on the Island, as it does in spring, it's easy to think that wind must have sandblasted the landscape into shape. Actually, water is the real artist. The Colorado and Green Rivers, descending from northern mountains through the desert, have cut deeply into the sedimentary rocks of Canyonlands. Already they have carried off 175 million years of geologic history in less than 10 million years. The Colorado is joined by the Green River at the **Confluence**, then rages through 14-mile **Cataract Canyon**, goes slack in **Lake Powell**, and, held back by

dams and spent in irrigation, slows to a trickle at its destination, the Gulf of California.

Originally sluggish and sinuous, these great rivers became mighty forces as the 130,000-square-mile Colorado Plateau was pushed up between 60 and 15 million years ago, eventually reaching its mile-high elevation. Cutting down in their looping courses, the meandering rivers chopped canyons and left behind a host of water-carved phenomena. Best known is the huge hairpin bend at **Deadhorse Point**, technically an entrenched meander. Farther south is a rash of natural bridges, formed when the rivers broke through their meanders in places like **Rainbow Bridge in Glen Canyon National Recreation Area** and nearby **Natural**

Incised meanders (above) show where the Colorado River has cut down deeply into the rocks.

White Rim sandstone (right) forms a conspicuous ledge at Island in the Sky's Green River Overlook.

Bridges National Monument.

The wild card in Canyonlands is salt. Sandstone and limestone cliffs, ledgy siltstones and mudstones, and crumbly conglomerates all sit uneasily on a basement composed of deep, shifting salt beds aptly named the Paradox Formation. Some 300 million years ago, movement along faults buckled the land, raising the Uncompahgre Uplift in the vicinity of today's La Sal Mountains and creating the basin that is now Canyonlands. Throughout the rest of the Pennsylvanian, Permian, and Triassic periods, an intermittent sea covered the basin, depositing salt beds thousands of feet thick.

When the sea eventually withdrew, Canyonlands was covered by a coastal plain braided by rivers flowing from the Uncompahgre highlands to the sea. These rivers carried red sediments from the mountains and mixed them with white sand dunes to create the colorful, crossbedded rocks that eventually fractured and eroded into the Needles. By Triassic times (some 230 million years ago), erosion had reduced the Uncompahgre mountains to hills and shed the washed-away material – clays, silts, sands, and pebbles – into the basin where they made the Moenkopi and Chinle Formations.

As the climate dried, blowing sand replaced inland seas across vast areas of the West, building massive dunes in Sahara-like environments. Both the Wingate and Navajo sandstone formations were laid down in such conditions during the Jurassic Period and make up the massive cliffs of canyon country. As this overburden grew, the unstable Paradox salt at the bottom flowed from the pressure until, finding its route blocked by deep-seated faults, it was forced up along fractures, pushing the overlying layers into domes that cracked and collapsed. One way or another, the restlessness of salt created much of Canyonlands and nearby Arches National Parks.

Land of Arches

Arches abound in southern Utah's canyon country but find their greatest concentration in the northeast corner of the Paradox Basin beneath the La Sal Mountains, where more than 1,700 are protected within Arches National Park. Entry is easy; tearing yourself away is more difficult. Photogenic **Delicate Arch**, record-breaking **Landscape Arch**, spectacles-like **Double-O**, precarious **Balanced Rock** – here is a private world of standing stones and boulders, fins, keyholes, windows, arches, and spires, where the quiet transformation of the ordinary into the magical is just a raven's wingbeat away.

Salt, the protagonist in the 300-million-year-long drama of Canyonlands, is the silent partner behind the show here, too. But it had slightly different results with the younger, terracotta-colored Entrada Formation, which erosion has stripped from the neighboring park.

The **Salt** and **Cache Valleys** at the heart of Arches follow a predictable pattern: faulting, salt intrusion, doming, collapse, dissolution of sandstone by water, and widening of joints. This results in long chorus lines of fins and eventually arches.

Entrada Sandstone is prime arch-forming material. Its lower part is a soft siltstone that quickly erodes and undermines the overlying massive upper layer. Flaking begins at the bottom, then curves upward into an arch. As weathering continues, the arch grows ever more slender until it finally crashes to the ground.

The entrance to Arches is just off Highway 191, north of Moab. When you stop at the visitor center, you'll have a great view across the highway of the huge displacement along the fault that created Moab Valley. A paved drive winds through the park, with turnouts leading to key features. Allow time to go at least as far as the **Windows**, hiking as many of the easy trails as you can. The scenic drive ends at **Devils Garden Campground**, where a pleasant sandy trail leads to Landscape Arch.

Exploring on Land

Trails at Canyonlands lead to dramatic geological features and breathtaking vistas. At

Arches National Park (left) contains more than 1,700 of these beautiful erosional features.

A slot canyon (opposite, top) in the Needles has developed where erosion has enlarged the joints in the sandstone.

Molar Rock (opposite, bottom) is a knob of sandstone that shows how differences in rock texture can shape its weathering.

colored Chinle and Moenkopi rocks. Geologists once thought this was a collapsed salt dome. However, the late Eugene Shoemaker, who founded the U.S. Geological Survey's lunar geology section, concluded that there ought to be a large meteorite impact crater on the Colorado Plateau and theorized that this crater could well be it. It's possible, in fact, that the crater is both, with the impact fracturing a salt dome, causing it to collapse.

Many people prefer to use four-wheel-drive vehicles to get deeper into Canyonlands backcountry, but this type of travel is not for the uninitiated. Jeep trails are not roads but difficult routes over deep sand

Island in the Sky, an easy one-mile loop goes to **Mesa Arch**, which frames picture-perfect views of **Washer Woman Arch** and the La Sals. The trail passes through high-desert vegetation of pinyon and juniper, blackbrush, and other scrub that, despite the aridity, find footings in one of Canyonlands' most important features: cryptobiotic soil. The texture of dark brown sugar, this fragile soil is a marvel of adaptation, spun from strands of blue-green algae and moss that retain water and stop erosion while new soil is forming. It's vital not to step on these mounds, which may have been forming for 500 years, only to be destroyed in a moment of human carelessness. If you walk off-trail, stick to slickrock and sandy washes.

Just west of Willow Flat Campground is one of Canyonlands' geological mysteries. **Upheaval Dome** is filled with a jumble of upended White Rim sandstone and multi-

and bare rock "desert pavement" chinked with stones where washouts have caused potholes. Progress is slow, and the constant bumping hard on the kidneys. But if this seems like fun to you, then Canyonlands offers the ultimate four-wheeling experience.

Kent Frost, one of Canyonland's earliest travel guides and a local rancher's son, was the first to bring paying guests into the Needles and the Maze by Jeep. Today, you can rent Jeeps in Monticello and Moab or travel with one of several outfitters. There's a fine view of the Colorado River from a four-wheel-drive route

narrow slot between two fins, where even the gentlest footsteps echo loudly. On the other side, the trail passes through a pinyon-and-juniper "garden." Then it crosses the sandy Elephant Canyon wash and ascends into the grassy meadows and spires of the Needles, once used as grazing by local ranchers but now protected by the park.

To make a long day of it, continue on the **Joint Trail** and loop back to the trailhead or backtrack to Elephant Canyon and hike along the wash to 200-foot-high **Druid Arch**, discovered in 1959. For longer explorations, get a permit, reserve a backcountry campsite, and continue toward the river. Here, the action of collapsing salt becomes more obvious, particularly in the 300-foot-deep alleyways of the **Grabens**, where brittle sandstone has been yanked along by moving salt and massive blocks have dropped between faults. A hiking trail descends **Lower Red Lake Canyon** to the river, where you can see **Spanish Bottom** in the Maze on the opposite bank.

behind the **Needles Visitor Center** (with a challenging final mile to negotiate), but the goal of many drivers is to make it over **Elephant Hill** – definitely one for the books if you're game.

For hikers, the best trip in the Needles is seven-mile **Chesler Park Trail**. Starting at Elephant Hill, the trail ascends quickly into massive, jointed rocks, which crowd around you like gossipy pals. To the east are the Abajos and the distinctive twin forms of the Sixshooter Peaks beyond the park boundary. To the north, the cliffs of Island in the Sky float out of the haze like a Manhattan skyscraper. The trail moves down through a

Touring by River

The Colorado River offers a way into the remote **Maze** country. River trips are easily set up through outfitters in Moab who offer a variety of half-day floats and longer ones, including a multiday white-knuckle trip through **Cataract Canyon**. The put-in is just outside Moab at Potash; takeouts might be at a canyon below Island in the Sky (where you pick up a Jeep and slowly bump the

2,000 feet up to the rim on the Shafer Trail), near the Confluence at Spanish Bottom, or on the other side of Cataract Canyon in Glen Canyon.

River journeys offer a different perspective on the landscape, which from below seems hushed and cathedral-like, while alive with tamarisk, waterfowl, canyon wrens, and other life at water's edge. Boats glide past older formations, shelves and alcoves, and oddly eroded rocks (given fanciful names like Yogi Bear and Booboo by individual river guides). Mudstones laden with fossil crinoids and dark chunks of petrified wood line the riverbank.

Mesa Arch (opposite) at sunset. Erosion will eventually cause the span to collapse.

Two hikers (left) stand silhouetted in one of Canyonlands' rocky crevasses.

Upheaval Dome (below) appears to be the eroded remnant of an ancient impact crater, complicated by an underlying layer of salt.

With any luck, you'll be here when erosion and gravity reconfigure the land during a sudden rockfall. One such event happened near Deadhorse Point in 1997 after heavy El Niño rains saturated the rock. The fall sent a tsunami-like wave over the opposite bank. If, like a veteran river guide, your eyes light up with joy at such a prospect, then you'll know that Canyonlands has worked its magic on you, too. And you'll be back.

TRAVEL TIPS

DETAILS

When to Go

Spring and fall are the most pleasant seasons, though you may want to avoid the hordes of mountain bikers who swarm Moab and other small towns during spring break. Summer temperatures top 100°F at lower elevations, and thunderstorms occur regularly from July to September. Winter brings heavy snowfall and freezing temperatures.

How to Get There

Major commercial airlines serve Salt Lake City International Airport, about four hours northwest of Moab. Commuter airlines serve Moab Canyonlands Airport and Walker Field Airport in Grand Junction, Colorado.

Getting Around

A car is essential in Canyon Country. Car rentals are available at the airports. High-clearance four-wheel-drive vehicles are often necessary on unpaved country roads.

Backcountry Travel

Overnight trips (and some day trips) into Canyonlands National Park require backcountry use permits; call the park for details.

Handicapped Access

The visitor center and some campsites are accessible. Contact the park's handicapped access coordinator for details.

INFORMATION

Arches National Park

P.O. Box 907, Moab, UT 84532; tel: 435-259-8161.

Bureau of Land Management

Sand Flats Road, P.O. Box M, Moab, UT 84532; tel: 435-259-7012.

Canyonlands National Park

2282 Southwest Resource Boulevard, Moab, UT 84532-8000; tel: 435-259-7164.

Grand County Travel Council

P.O. Box 550, Moab, UT 84532; tel: 435-259-8825 or 800-635-6622.

CAMPING

Campsites are available at both Canyonlands and Arches National Parks on a first-come, first-served basis. Reservations are accepted at nearby Deadhorse State Park, 800-322-3770, up to 16 weeks in advance of arrival.

LODGING

PRICE GUIDE – double occupancy

$ = up to $49 $$ = $50–$99
$$$ = $100–$149 $$$$ = $150+

Aarchway Inn

1551 North Highway 191, Moab, UT 84532; tel: 800-341-9359 or 435-259-2599.

This new, two-story motel offers elegant accommodations amid the red-rock cliffs and buttes of the Colorado River, two miles from Arches National Park. The inn has 97 rooms and suites, all with queen- or king-sized beds, refrigerators, microwaves, and wet bars. An outdoor swimming pool, indoor hot tub, and exercise facility are available. A complimentary continental breakfast is served. $–$$$$

Castle Valley Inn

HC 64, Box 2602, Castle Valley, UT 84532; tel: 435-259-6012.

This stone-and-wood inn on La Sal Loop Road is surrounded by soaring red-rock cliffs near Moab and the Colorado River.

The inn has five guest rooms with private baths in the main building, and three bungalows with kitchenettes and barbecue grills. Rooms are decorated with Navajo, Hopi, and Zuni Indian furnishings. $$–$$$$

Mayor's House Bed-and-Breakfast

505 East Rose Tree Lane, Moab, UT 84532; tel: 888-791-2345 or 435-259-6015.

The Mayor's House has six guest rooms and suites, with queen- and king-sized beds, whirlpool tubs, and VCRs (a movie library is available). Two rooms share a bath. Built in the 1980s, the inn features Southwestern decor. A heated swimming pool, hot tub, and barbecue patio are on the premises. $$–$$$$

Pack Creek Ranch

P.O. Box 1270, Moab, UT 84532; tel: 435-259-5505.

A rustic retreat on the La Sal Loop Road near Moab, this popular guest ranch offers 12 cabins, fine dining, massage therapy, hot tub, sauna, swimming pool, horseback riding, pack trips, and other resort-style amenities. $$–$$$

Redstone Inn

535 South Main Street, Moab, UT 84532; tel: 800-772-1972 or 435-259-3500.

This rustic-looking motel, centrally located, offers 50 rooms with log furniture, refrigerators, and microwaves. A picnic area, gas barbecue grill, and 24-hour laundry facility are on the premises. Guests have use of the swimming pool at nearby Big Horn Lodge. $–$$

Sunflower Hill Bed-and-Breakfast Inn

185 North 300 East, Moab, UT 84532; tel: 435-259-2974.

This country-style bed-and-breakfast occupies a garden setting three blocks from downtown Moab. The inn has 11 guest rooms with antique beds and private baths. Some rooms have whirlpool tubs, private balconies, garden patios, and VCRs. A hot tub, library, and laundry facility are on the premises. A large home-cooked breakfast is served. Winter rates available. $$–$$$

TOURS

Adrift Adventures

378 North Main Street, P.O. Box 577, Moab, UT 84532; tel: 800-874-4483 or 435-259-8594.

The company leads jetboat and Jeep tours between Moab and Canyonlands, whitewater rafting on the Colorado River, and rafting and horseback riding excursions at the base of the La Sal Mountains.

Lin Ottinger's Tours

Moab Rock Shop, 600 North Main Street, Moab, UT 84532; tel: 435-259-7312.

Known as the Dinosaur Man (he even has a dinosaur named after him), geologist and naturalist Lin Ottinger has been offering geological and paleontological tours of Canyonlands and Arches National Parks since 1960.

Tag-A-Long Expeditions

452 North Main Street, Moab, UT 84532; tel: 800-453-3292.

The outfitter offers four-wheel-drive and jetboat trips on the Colorado River in Canyonlands National Park. Guided trips explore the area's natural, geological, and cultural history. A river-running day trip explores Westwater Canyon, the heart of the ancestral Rockies, made of metamorphic gneiss.

MUSEUMS

Dan O. Laurie Museum

118 East Center Street, Moab, UT 84532; tel: 435-259-7985.

Museum exhibits examine geology, archaeology, and the pioneer history of southeast Utah.

Excursions

Capitol Reef National Park

HC 70, Box 15, Torrey, UT 84775; tel: 435-425-3791.

Capitol Reef, in south-central Utah, preserves the 15-mile-wide, 100-mile-long Waterpocket Fold, a flexure in the Earth's crust that has been eroded into narrow canyons topped with giant domes of creamy Navajo sandstone. Paved Highway 24 crosses the northern portion of the Fold beneath soaring cliffs of Wingate sandstone along the verdant Fremont River. The southern section offers a geological thrill ride over the Burr Trail, an old rancher's route, and down colorful 1,000-foot "breakers" of upended Wingate, Kayenta, and Navajo rock.

Great Sand Dunes National Monument

11999 Highway 150, Mosca, CO 81146; tel: 719-378-2312.

Sand dunes 700 feet high, the tallest in America, can be found a few hours from the Four Corners, in southern Colorado's San Luis Valley. Southwesterly winds reaching 40 miles per hour scoop up sand shed from the San Juan Mountains and drop it at the base of the Sangre de Cristo Mountains. The shifting dunes are most photogenic at dawn or dusk. Trails lead through different life zones in the dunes, along Mosca Creek, and on Mosca Pass.

Natural Bridges National Monument

P.O. Box 1, Lake Powell, UT 84533; tel: 435-692-1234.

Often overlooked, this park in southeastern Utah is the site of three sandstone bridges carved by tributaries of the Colorado River. Sipapu Bridge, at a height of 220 feet, is the second largest ever recorded. (Nearby Rainbow Bridge on Lake Powell is the largest.) Hike the nine-mile loop past the bridges and watch for prehistoric Indian dwellings, rock art, and wildlife. A delightful campground offers sandy tent pads, mesa-top views, and interesting campfire talks by rangers.

Meteor Crater
Arizona

CHAPTER 13

From the rim of **Meteor Crater**, a vast empty space 4,100 feet across lies at your feet. The upper wall drops away in a treacherous slope that gradually flattens to meet the bowl-shaped bottom. Voices mingle with the wind as breezes warmed by sunlit rock blow past you into the huge natural amphitheater. This eerie place is where nearly a cubic kilometer of rock disappeared in a blinding flash. ◆ It could well be called the Rosetta Stone of astrogeology. Studies of the crater opened up planetary scientists' understanding of craters on Earth, the Moon, and throughout the solar system. It was Meteor Crater that led scientists to an astonishing fact: Impact craters such as this are by far the most common landform in the solar system. That's a finding the rest of humanity has yet to absorb. ◆ Meteor Crater lies about six miles south of Interstate 40 between **Flagstaff** and **Winslow**, Arizona. Driving east from Flagstaff on I-40, you get your first clear view of the crater at milepost 226.

A huge piece of iron fell from space onto the Colorado Plateau, resulting in the world's best-preserved impact crater.

Pull the car onto the shoulder and get out. You'll see the crater's elevated rim breaking the horizon to the southeast – a long, raised hillock, tan and rumpled, about as wide as two fingers held at arm's length. From where you stand, the rim is 11 miles away. ◆ The day the meteorite struck 50,000 years ago, the land looked much as it does now. From here you'd have seen a bright finger of light high in the southeastern sky. It angled about 45 degrees from vertical as it traveled toward you. The finger, bright as a second Sun, touched down with a brilliant, soundless flash. The noise of impact and the blast wave took a minute to reach your spot, rolling on like an enormous thunderclap, with the blast raising dust as it came.

Meteor Crater has slowly filled with debris washed down the walls or blown in by the wind. Moments after impact, the floor was about twice as deep as it is now.

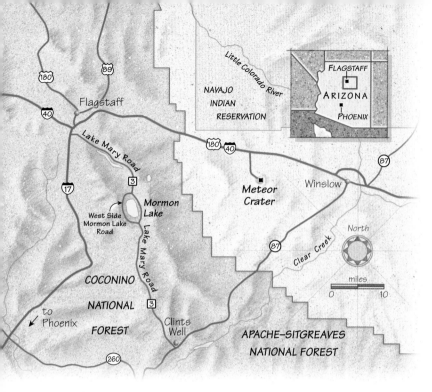

In the distance, a tall cloud column rose out of the dusty haze, mushrooming at the top. Around the base was a denser cloud kicked up by flying debris as layers of rock were overturned and flung outward. The crater's surroundings were hammered by fragments the size of automobiles falling from the cloud. Minutes passed, the dust settled, and the mushroom cloud drifted clear. Half an hour after impact, everything looked pretty much as before – except for the raised crater rim and the hummocky blanket of ejecta around it.

Celestial Iron Mine?

The crater is forever linked to one man's name – indeed, his family still owns it. Daniel Moreau Barringer (1860-1929) was a mining engineer with rare energy. Hearing of the crater and nickel-iron meteorites that were found nearby, Barringer concluded that the crater had been produced by an impact. He figured that the meteorite was still there underground, and he hoped to mine it for tons of iron and nickel.

After staking claim to the land, Barringer began by drilling a few holes in the crater's center. As the bit worked deeper, it penetrated hundreds of feet of rock flour (shattered quartz), then broken rock, and finally undisturbed sandstone. Barringer found no big meteorite but instead countless tiny nickel-iron fragments scattered like poppy seeds in a cake. Refusing to give up, he spent one borrowed fortune after another drilling dozens of holes – always seeking the main mass of the meteorite, always with the same null results.

In 1929 astronomer Forest Ray Moulton published a report that doomed all hope of finding a large intact meteorite. Moulton calculated the energy that the falling meteorite would have delivered and concluded it was much smaller than Barringer had assumed. Moreover, the meteorite had been largely destroyed in the impact. Barringer, bitterly disappointed, died soon after.

Moulton's report got astronomers thinking seriously about impacts and helped put crater studies on the scientific map. Over the years, the best return for the hundreds of thousands of dollars that Barringer spent on drilling has come from tourists drawn by the crater's fame. Moon-bound astronauts trained here, and today it's a scientific mecca, the source of numerous papers published every year. After all, this is the best-preserved impact crater on the planet.

Twisted Walls

At mile 233, Interstate 40 meets the road leading south to the crater, whose uplifted rim is visible most of the way. The road weaves around reddish brown outcrops of Moenkopi sandstone that look like big stone biscuits. At 4.2 miles, the road crests one last ridge. The crater rim is now a wide hill that dominates your view, with blocks of debris strewn across the landscape. Coming down from the ridge, the road turns east and ascends over the ejecta to the parking lot.

Your first look into the crater is dramatic. As you stand on the north rim, the south wall is about two-thirds of a mile away. The floor, over 500 feet below, consists of geologically recent alluvium, material washed off the walls and blown in by the wind.

Four rock layers are exposed in the walls. Lowest and oldest is the Permian age Coconino sandstone, grayish white in color and about 260 million years old. Only the Coconino's upper part is seen in the crater. This fine-grained rock with a sugary texture originated as a vast field of sand dunes.

The Coconino is overlain by the Toroweap Formation, a Permian sandstone mixed with dolomite. Yellowish in color and about five feet thick, it formed under relatively shallow water, showing that the ocean crept in and buried the Coconino dunes, probably while they were still loose sand.

The next higher layer is the Kaibab Formation, cream colored and Permian in age (250 million years). This 260-foot-thick unit, made of dolomite and limestone, is another oceanic deposit, though the water had deepened since the days of the Toroweap. The Kaibab is the rock layer in the clifflike section in the crater's wall near the top.

The crater's upper rock layer is the reddish brown Moenkopi Formation, from the Triassic age, about 240 million years old, and averaging 30 feet thick. It is a thickly bedded and massive sandstone in its lowest part, but changes by the top to a thin-layered

Petrified logs (below) lay jumbled in a gully at Petrified Forest National Park. Living wood becomes petrified (right) when a tree is buried and its cellulose tissues are slowly replaced with silica carried in groundwater.

Trees of Stone

Seventy miles east of Meteor Crater lies beautiful **Crystal Forest**. Chips of petrified wood sparkle in the Arizona sunshine. Entire tree trunks lay prostrate on the ground, fractured so cleanly they seem to have been bucked up by a chainsaw. So perfectly are they preserved that the bark and heartwood appear fresh and alive. A closer look reveals that this isn't freshly cut timber but trees that have turned to stone.

The trees in Petrified Forest National Park were once alive – in the Triassic age. What is now the high, dry desert of northern Arizona hovered down by the equator 225 million years ago. Tall *Araucarioxylon*, *Woodworthia*, and *Schilderia* trees thrived in the tropical environment. The trees, blown down in storms, were washed into streams and rivers, then buried in mud and volcanic ash, where they slowly decayed. Their woody tissues were replaced by silica transported in groundwater – turning the wood into jewel-colored jasper, agate, and quartz.

The volcanic ash and dust that engulfed the trees accumulated in the Chinle Formation. The eroded Chinle hills, called badlands, are nothing more than big mud piles in garish stripes of orange, blue, gray, chocolate, lavender, and cream. The colors have given name to the **Painted Desert**, which makes up an enthralling wilderness in the park's northern part. Rich in bentonite clays, the Chinle shrinks and swells with every wetting and drying cycle, giving it the texture of elephant skin. Water and wind melt away the soft muds, revealing the wealth of petrified wood.

siltstone. The sandstone was laid down in relatively shallow water, but the siltstone has ripple marks indicating that the ocean environment had grown quite shallow, with widespread mudflats.

The brown Moenkopi, though not visible everywhere, makes a good marker for locating the pre-impact land surface. It is thin and usually lies only a short distance below the rim crest. In places it appears thick, but this is where Moenkopi debris has trickled down over the topmost Kaibab layers and stained them. As a general rule, the cliffs are Kaibab, while the Coconino lies buried under the loose slope of talus below. The rim materials above the Moenkopi are mostly layers of Kaibab overturned in the impact. (The museum itself rests on these overturned beds.)

Spend a moment following layers in the walls with binoculars. Gullies have developed on the upper wall where offset strata indicate faults. Examine the Moenkopi and Kaibab layers on the east and west. See how they tilt? Those beds were horizontal before the impact, but the explosion – equal to 20 megatons – heaved them up, and they now dip outward. For one or two seconds after the meteorite struck, shock waves caused the rocks to flow like toothpaste. When the energy of the intense shock dissipated, the deformed beds froze in place.

Meteor Crater Stratigraphy

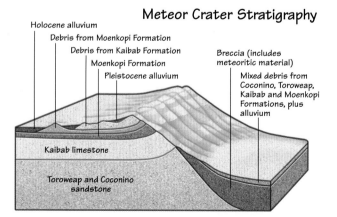

Holocene alluvium
Debris from Moenkopi Formation
Debris from Kaibab Formation
Moenkopi Formation
Pleistocene alluvium
Breccia (includes meteoritic material)
Mixed debris from Coconino, Toroweap, Kaibab and Moenkopi Formations, plus alluvium
Kaibab limestone
Toroweap and Coconino sandstone

Rim Walk

Guided tours of the rim leave the visitor center every hour between 9 A.M. and 2 P.M. (weather permitting)

The Dinosaur Diamond

Barringer Point (left) is the highest part of the rim tour and provides a good view into the crater.

The killer claw (right) of *Utahraptor* sliced open its prey; the fleet-footed predators may have hunted in packs.

The skull of *Camarasaurus* (bottom) emerges from the nearly vertical bone bed at Dinosaur National Monument.

Who can forget velociraptor, the swift, predatory dinosaur of *Jurassic Park*? Traveling in a pack and able to run upright, it brought down prey far larger than itself with ease, then killed with one fatal slash of its enormous claws.

Fiction, you say? Well, hold onto your hat. Even as the movie began production, Jim Kirkland and fellow paleontologist Carl Limoni were at work in a remote dinosaur quarry north of Moab, Utah. Uncovering what they thought was an ankylosaur bone, they quickly realized they had found a new species altogether. *Utahraptor*, as it was eventually called, was velociraptor personified.

Such finds are rare, but the largely Triassic, Jurassic, and Cretaceous rock strata of the Colorado Plateau make a great place to search for fossils (though fossil-collecting is forbidden on public lands). In fact, some of the oldest dinosaurs ever found were unearthed just a short drive from Meteor Crater at Petrified Forest National Park.

Other sites in the bone-rich Dinosaur Diamond, as the region is sometimes called, include **Dinosaur National Monument** on the Utah-Colorado border, where some 1,600 fossils have been left exposed on a quarry wall. To the south, in Colorado's Grand Valley, **Mygatt-Moore Quarry** has produced about 2,000 bones from eight dinosaurs since its discovery in 1981. *Supersaurus*, one of the largest dinosaurs found to date, was discovered in 1972 outside of Montrose, Colorado, at **Dry Mesa Quarry**; visitors can watch the excavations in summer. The infamous *Utahraptor* will eventually be displayed at the **College of Eastern Utah Prehistoric Museum** in Price, Utah. The museum displays dinosaur eggs and some of the 12,000 bones taken from nearby **Cleveland-Lloyd Dinosaur Quarry**, one of the world's primary sources of allosaur skeletons.
– *Nicky Leach*

and go about one-third of a mile to the west. Visitors are not allowed to wander freely along the rim, to prevent accidents and also to preserve the scientific integrity of the site.

On the rim trail, you pass through a broad low notch in the northern wall. A mule path heads down through here, made to haul Barringer's drilling equipment to the crater floor. The trail leads to the remains of the old museum, destroyed by fire in the 1950s. It was constructed from brown blocks of Moenkopi sandstone, some showing ripple marks. The old museum once housed the largest piece of the meteorite yet found, the 1,406-lb Holsinger fragment, now on display in the new museum, along with excellent exhibits on how craters form and their role in sculpting the face of the solar system.

As you walk the trail, use your binoculars to explore farther. Try to imagine the unbelievable violence that tossed around so much rock so casually. Look at the pieces underfoot. If you find grayish Coconino sandstone, rub it with your fingers. Feel the sugary texture? With a hand lens, examine the large and sparkly grains. Each grain spent uncounted ages knocking against others and losing its sharp edges. Let your imagination go back millions of years to an ancient field of dunes spreading for thousands of square miles.

Away from the throng of visitors, the silence of Meteor Crater speaks most strongly. This haunting place tells of times long past, when its rocks were formed, and also of a time much more recent – when a piece of the sky fell to Earth and brought the solar system home.

TRAVEL TIPS

DETAILS

When to Go

Summer is sunny and warm, with highs in the 80s. Thunderstorms are common in July and August. Winter is cold but often sunny, with daytime temperatures in the 40s. Annual snowfall averages 100 inches.

How to Get There

Major airlines serve Phoenix, Arizona, 145 miles away, and Albuquerque, New Mexico, 325 miles. America West Airlines serves Flagstaff.

Getting Around

A car is the only convenient way to visit the site. Rentals are available at the airports.

Handicapped Access

Elevators and lifts provide access to the museum, gift shop, and observation area immediately behind the museum. The rim tour is not accessible.

INFORMATION

Meteor Crater Enterprises
P.O. Box 70, Flagstaff, AZ 86002; tel: 800-289-5898 or 520-289-5898.

Flagstaff Visitor Center
1 East Route 66, Flagstaff, AZ 86001; tel: 520-774-9541.

CAMPING

Bonito Campground
Peaks Ranger Station, 5075 North Highway 89, Flagstaff, AZ 86004; tel: 520-526-0866.

This campground, at Sunset Crater National Monument 18 miles northeast of Flagstaff, has 44 sites and flush toilets, but no showers. It is open April through October.

Meteor Crater RV Park
Off Highway 40 at Meteor Crater, P.O. Box 70, Flagstaff, AZ 86002; tel: 520-289-4002.

This park, owned by Meteor Crater Enterprises, has 71 drive-in sites. It also has restrooms and shower facilities, including two wheelchair-accessible restrooms and showers. The park also offers a recreation room, playground, laundry, store, and gas station.

LODGING

PRICE GUIDE – double occupancy

$ = up to $49	$$ = $50–$99
$$$ = $100–$149	$$$$ = $150+

Best Western Pony Soldier
3030 East Route 66, Flagstaff, AZ 86004; tel: 520-526-2388.

The two-story chain motel just outside of downtown Flagstaff offers 90 comfortable rooms, a heated indoor pool, and restaurant. $$

Fray Marcos Hotel
1201 West Route 66, Suite 200, Flagstaff, AZ 86001; tel: 800-843-8724.

A reconstruction of the original 1908 Fray Marcos, this new 89-room hotel at the historic depot mixes turn-of-the-century ambiance with modern amenities. The lobby and saloon are decorated in a style reminiscent of the Old West. $$–$$$$

Inn at 410
410 North Leroux Street, Flagstaff, AZ 86001; tel: 520-774-0088 or 800-774-2008.

This bed-and-breakfast has one single room and eight two-bedroom suites. Situated about four blocks north of downtown Flagstaff, the house was built in

1907 and is furnished in period antiques. Rooms have air conditioners and refrigerators, but no telephones or televisions. $$$–$$$$

Lake Mary Bed-and-Breakfast
5470 South J Diamond Road, Flagstaff, AZ 86001; tel: 520-779-7054 or 888-241-9550.

This inn on the southern edge of Lake Mary has four rooms with private baths – two with tubs, two with showers. The house was built in 1930 in Jerome, Arizona, and was later moved to this spot outside of Flagstaff. A restaurant is nearby. $$–$$$

Little America Hotel
2515 East Butler Avenue, Flagstaff, AZ 86004; tel: 520-779-2741.

This large hotel east of downtown Flagstaff has 247 rooms in several buildings on 500 acres of tree-shaded grounds. Amenities include a dining room, coffee shop, heated pool, Jacuzzi, and jogging trail. $$–$$$

TOURS

Meteor Crater Enterprises
P.O. Box 70, Flagstaff, AZ 86002; tel: 800-289-5898 or 520-289-5898.

Daily guided tours go about a third of a mile around the crater rim from the museum. Wear adequate hiking shoes. The tours run from 9 A.M. to mid-afternoon and last about an hour, weather permitting.

MUSEUMS

Lowell Observatory
1400 West Mars Hill Road, Flagstaff, AZ 86001; tel: 520-774-2096.

The world-famous observatory's main focus is on astronomy and other planets (including Pluto, which was discovered here in 1930). Yet the displays in its visitor center draw links between the broad field of planetary science and the particular planet on which we happen to live. The result is a much richer picture of Earth than geology alone provides.

Museum of Astrogeology

Meteor Crater Enterprises, P.O. Box 70, Flagstaff, AZ 86002; tel: 800-289-5898 or 520-289-5898.

The museum features exhibits on meteoritics and the mechanics of impact cratering. Interactive computer displays examine meteorites and asteroids, the solar system, and Comet Shoemaker-Levy 9, which struck Jupiter in 1994. An 80-seat movie theater shows *Collisions and Impacts* twice each hour. Also on display is a spacesuit worn on the Moon by Apollo 12 crewman Charles Duke.

Museum of Northern Arizona

3101 North Fort Valley Road, Flagstaff, AZ 86001; tel: 520-774-5213.

This excellent museum focuses on the geology, biology, paleontology, and archaeology of the Colorado Plateau.

Walnut Canyon National Monument

Walnut Canyon Road #3, Flagstaff, AZ 86004; tel: 520-526-3367.

Although this site is primarily archaeological, two self-guided trails lead visitors to a 350-foot-deep gorge in a bed of Kaibab limestone. Observant hikers may find fossils, mainly brachiopods, in the walls along the trail.

Excursions

El Malpais National Monument

123 Roosevelt Avenue, Grants, NM 87020; tel: 505-783-4774.

Southeast from Grants lies a skein of lava flows so tangled that it's virtually impassible. "The badland" – *el malpaís* in Spanish – includes cinder cones, lava tubes (including some with year-round ice), and lava squeeze-ups. The Park Service visitor center is on Highway 53, 23 miles south of Interstate 40; a new multi-agency visitor center is in Grants (1900 East Santa Fe Avenue; tel: 505-876-2784).

Sunset Crater National Monument

Route 3, Box 149, Flagstaff, AZ 86004; tel: 520-526-0502.

Sunset Crater, about 15 miles northeast of Flagstaff, is one of Arizona's youngest geologic features. The volcano was born in A.D. 1064 or 1065, and eruptions continued to belch forth for about 200 years, building a cinder cone that now stands roughly a thousand feet high. Gases stained the cinders, creating the "sunset" colors for which the volcano is named. Visitors are  not allowed to climb Sunset Crater, but a nature trail winds across a rugged lava flow at its base and approaches the foot of the cone. A second, steep trail leads to nearby Lenox Crater.

Mogollon Rim

Arizona Office of Tourism, 1100 West Washington Street, Phoenix, AZ 85007; tel: 800-842-8257 or 602-542-8687.

The Mogollon (muggy-OWN) Rim is an escarpment that forms the Colorado Plateau's southwestern edge as it winds across north-central Arizona. Composed mostly of sedimentary rocks but overlain in places by volcanic flows, the rim marks the boundary between the hot deserts of the south and the cool northern highlands of Ponderosa pine. Drivers can climb the rim on Highway 89A from Sedona to Flagstaff through spectacular Oak Creek Canyon.

Grand Canyon
National Park
Arizona

CHAPTER **14**

The Grand Canyon stretches the limits of comprehension. Standing on the rim and gazing into the void, we have little to put the canyon into human perspective. It is immense: more than a million acres in the national park, close to 6,000 feet deep from the North Rim to the bottom, 10 miles across as the raven flies, 277 miles of the Colorado River coursing across northern Arizona. And its rocks are very, very old: 1.8 billion years of Earth's history are showcased within its walls. ◆ A walk down a trail is really the only way to grasp this great wonder – and the **South Kaibab Trail**, which starts at **Yaki Point** on the **South Rim**, is a good place to begin. The hike is nearly seven miles from the rim to the Colorado River, seven steep, knee-punishing miles. Most hikers make a beeline for the canyon bottom, failing to pause long enough to **Plumb the depths of** examine the wonders along the way. Too bad. For **geologic time with** the South Kaibab Trail passes through a cross **a journey into the** section of nearly all the strata in the canyon, arranged **heart of the Earth.** mostly in neat, horizontal bands that obligingly obey the geologic rule of younger rock on top of older. With a few notable exceptions, almost all are sedimentary rocks, laid down by water and wind. ◆ Strap on a pack and fill the water jugs (a gallon a day per person is the minimum recommendation, especially on the South Kaibab Trail, where no water is available until the river). Then drop below the rim into another world of time and space. The trail switchbacks quickly down through the Kaibab and Toroweap Formations. The cream-colored Kaibab is chock-full of fossil sponges and the remains of other creatures that lived in the ocean that inundated this part of the world in Permian times. The Kaibab, the caprock on both the South

The Colorado River flows below Toroweap Point. The river has carved a canyon a mile deep in less than six million years.

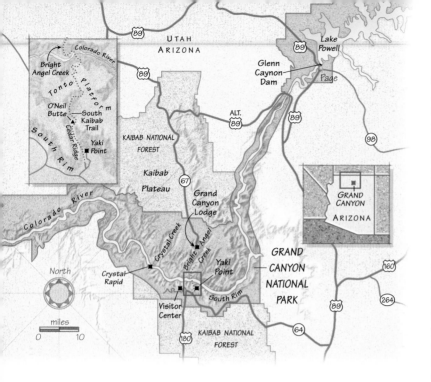

share its marine heritage. A clean line – or contact – separates the Toroweap from the strikingly different Coconino sandstone underneath. The Coconino is a nearly pure quartz sandstone, golden colored, with beautiful crossbeds. These mark the direction of the wind that laid up dunes in a vast ancient desert – a wind frozen in rock. The Coconino forms a solid line of cliffs 300 feet high in places, permitting passage only where faults have broken through this formidable stone barrier.

and North Rims, is about 250 million years old – the youngest rock in the canyon. Still, it is older than dinosaurs, mammals, and flowering plants.

The limestones and sandstones of the Toroweap, just slightly older than the Kaibab,

Down through the Coconino, the trail emerges onto a flat clearing called **Cedar Ridge**, a good place to stop and watch slender lizards scamper from rock to shrub, while ravens croak and perform their aerial ballets. The gaudy brick-red rock is Hermit shale, the color due to small doses of hematite. The Southwest is rich in such redbeds that resulted when the rocks were laid open to the air and iron pigments oxidized: the rocks, in essence, rusted. Impressions of fern leaves and dragonfly wings speak of the swamp and lagoon environment in which the Hermit was deposited 280 million years ago.

Below Cedar Ridge, the trail swings around **O'Neill Butte** and steps down through the Supai Formation, a group of siltstones and sandstones also produced in a low, swampy environment. When wet, the Supai turns to a sticky clay that deters a hiker's progress.

The trail then dives through a break in the massive Redwall limestone, another impressive wall of cliffs girding the canyon's midsection. Fossil corals, crinoids, brachiopods, and bryozoans are plentiful in the Redwall, typical of the creatures that reigned in the warm, clear ocean in which it was deposited. The Redwall, incidentally, is actually gray rather than red; it is stained by pigments leaching off the Supai above.

Havasu Creek (opposite), a tributary of the Colorado River, attracts explorers.

Pima Point (left) affords a spectacular view of the canyon's stratigraphy, from the dark Vishnu schist of the Inner Gorge to the Kaibab limestones of the distant North Rim.

"Fluted" limestone (above) is the product of weathering.

The Depths of Time

By now, hikers find themselves in true desert. The canyon's profound silence turns voices to whispers. And by now legs are feeling the brunt of the unrelenting downhill passage. The wide-open **Tonto Platform** is a monotone of muted grays and greens, with great swaths of low, widely spaced shrubs called black-brush, lending the air of an oriental garden to this part of the canyon.

The next three major rock layers are the Muav limestone, Bright Angel shale, and Tapeats sandstone. The trio reflects the grada-tion of environments of the 530- to 550-million-year-old Cambrian sea in which they formed: the Muav is the carbonate layer of the deepest water, the Bright Angel the offshore muds, and the Tapeats the onshore beach. A look at an ocean today reveals the same processes at work. The present, geologists say, is the key to the past.

Onward and deeper. Underlying these are the vivid orange and brown beds of the Grand Canyon Supergroup. This is a thick sandwich of interbedded sedimentary

Grand Canyon Stratigraphy

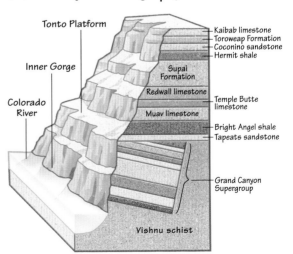

Tonto Platform

Inner Gorge

Colorado River

Supai Formation

Redwall limestone

Muav limestone

Kaibab limestone
Toroweap Formation
Coconino sandstone
Hermit shale

Temple Butte limestone

Bright Angel shale
Tapeats sandstone

Grand Canyon Supergroup

Vishnu schist

John Wesley Powell (right) confers with his Paiute guide; Powell lost his right arm at the Battle of Shiloh.

John Wesley Powell

In outlining his plan to go down the Colorado River by boat, John Wesley Powell proclaimed that "the Grand Canyon of the Colorado will give the best geological section on the continent."

The adventurous, self-trained geologist felt confident that this unexplored region would reveal its mysteries to him. In May 1869, he and his crew of nine seasoned men set out in four wooden boats from Green River, Wyoming. Three months later they entered the Grand Canyon. As they journeyed, Powell wrote, "The book is open, and I can read as I run."

At the end of August that year, exhausted but exhilarated, Powell and crew completed the momentous expedition. Powell's journal of the trip contributed what historian Stephen Pyne considers "the classic expression of the view from the river, the words by which his generation appreciated its revelation, the images by which tourists throughout the 20th century have understood it."

A hero for his efforts, John Wesley Powell returned to make another successful run in 1870–71. He then entered the political and bureaucratic fray in Washington, D.C., founding and directing the U.S. Geological Survey and the American Bureau of Ethnology. From his canyon experience, he forged a plan for settlement of the West along watershed boundaries. But Powell's grand vision, based on science and experience, was stymied by politics, and he died a frustrated man.

and volcanic rocks, laid down during the late Precambrian, 1.2 billion to 800 million years ago. The layers were then lifted up, tilted, and beveled off, leaving exposed the remnants of what was once an entire mountain range. Observant hikers may note circular structures called stromatolites in this rock; these are algae deposits, vestiges of early life on Earth. The South Kaibab Trail is one of the few spots in the canyon where the Supergroup is exposed. Elsewhere, the layers eroded away completely, and the tan horizontal Tapeats sandstone was deposited directly on top of the upturned dark rocks of the **Inner Gorge**.

The Inner Gorge is composed of Vishnu schist – crystalline, black rock glistening with garnets and mica and beautifully ribboned by intrusions of pink granite. The Vishnu approaches two billion years in age. It is the oldest rock in the canyon, metamorphosed

from an earlier form under heat and pressure so great that had anything left fossil remains, they would have been eradicated. In summer, when the mercury tops 100°F, the schist becomes almost too hot to touch. Here, the canyon rims do indeed seem billions of years away.

The trail winds down another mile or so until finally, and mercifully, it crosses the bridge suspended high above the swirling waters of the Colorado River. From the rim, the river appeared as a tiny stream. But from this vantage point, its true size and power are obvious. The Colorado River, of course, is what made – if such a prosaic word is permissible – the Grand Canyon.

Sawing through the Rock

The Grand Canyon is a product of a river, a hill, and time. The "hill" – the **Kaibab Plateau** – poses a perplexing geological problem. Once the rocks were laid down, they were raised up, part of a wholesale uplift of the region known as the Colorado Plateau. The Kaibab is a

wrinkle on the Colorado Plateau; standing more than 9,000 feet high in places, it qualifies as a big, flat-lying mountain. Most self-respecting rivers go around a mountain rather than expend the energy to cut through it. But the contrary Colorado River did just the opposite, incising the Grand Canyon directly through the mountain.

Nineteenth-century explorer John Wesley Powell hypothesized that the river was

A sunset vista from Grandview Point (left) overlooks the buttes and mesas of the eastern side of the park.

Cape Royal (right) marks the southern tip of the North Rim's Walhalla Plateau.

The mouth of Redwall Cavern (above) frames the Colorado River and canyon walls.

Deer Creek Narrows (right) is one of the many deep, narrow side canyons that feeds into the main canyon.

there first and the land lifted up against its downcutting – like a knife slicing down through the rising layers of a cake. Others say the plateau was in place, and the river became entrenched within as it cut down. Whatever the truth, the Colorado River did the job in an amazingly brief period. The canyon itself – as opposed to the rocks exposed in its walls – is young, probably cut to its present depth in a mere two to six million years – scarcely a tick of the geologic clock.

It wasn't just flowing water, however, that did the carving. The Rio Colorado once carried millions of tons of sediments in its waters. That heavy load of sand, gravel, and large boulders acted like a giant rasp, grinding against the streambed, abrading the rock, and carrying it bit by bit downstream to the delta at the Gulf of California.

As the Colorado was carving the canyon's depth, a hundred or so side canyons ate back into the rims and widened the canyon. Erosion in the side canyons often proceeds at a slow, almost invisible pace. But at times, it can be sudden and spectacular. Slurries of mud, water, and rock called debris-flows roar down the drainages during flash floods. In **Crystal Creek** one December night in 1966, a winter storm delivered a sudden pulse of precipitation to the North Rim. Boulders that hurtled down during that event clogged the river channel and created the fearsome **Crystal Rapid**, with which river runners must now contend.

The erosive power of the Colorado River has been subdued in modern times by **Glen Canyon Dam**, fifteen miles above the Grand Canyon. The gates of the dam were closed in 1963, creating **Lake Powell**, which stores water and harnesses the river for hydroelectric power as well as recreation. The Colorado, once declared "too thick to drink and too thin to plow," now runs cold and clear most of the year. Sediments are held behind the dam, and the river's cutting action through the resistant schist has been greatly diminished

But this is not the first time the Colorado has been dammed. Nature has done it in

Mooney Falls (above) in Havasu Canyon lies within the Mississippian-age Redwall limestone.

West Mitten (right), like other buttes in Monument Valley, is composed of Permian-age DeChelly sandstone.

Monument Valley

The Navajo call Monument Valley *Tsé Bií Ndzisgaii*, meaning "place between the rocks." It's an intuitive description, for the geologic story of Monument Valley is very much a story of what's missing – of rock eroded away, of left-behind stone monoliths standing majestically amid wide-open spaces.

The bulk of the monuments are of DeChelly (pronounced de-SHAY) sandstone. It is relatively pure quartz, the end result of a thick accumulation of sand dunes that covered the region more than 200 million years ago. Because of its uniform makeup, the DeChelly tends to erode as cliffs. Also, underlying it is a softer, slope-forming claystone called the Organ Rock. Both are Permian in age – 280 to 240 million years old – corresponding to the uppermost layers of the Grand Canyon.

The hands of wind and water have been at work on the sandstone, carving elegant spires, arches, buttes, and mesas that have picked up names both literal and fanciful: **Full Moon Arch**, **Raingod Mesa**, **Totem Pole**, and **Ear of the Wind**. To understand how the "monuments" formed, imagine that at one time the pieces were all connected. Water, frost, plant roots, and other agents wheedled their way into minute cracks in the rock, forcing them apart. Ages of rain and wind finished the job, carving the mesas into buttes and spires. Grain by grain, pebble by pebble, over millions of years the intervening rock has weathered away and washed into the San Juan River. What remains is one of the signature landscapes of the American West.

A backpacker on the Kaibab Trail (left) takes in a glorious view of mesas and temples marching toward the North Rim.

Columnar basalt formations (below) mark one of the lava flows that cascaded into the canyon and temporarily dammed the river.

A hiker climbs into narrow Matkatamiba Canyon (opposite).

a big way at least a dozen times over the last million years. In the western end of the canyon, volcanoes erupted, lava cascaded over the rim and sizzled into the river, obstructing its course and backing up water 175 miles or more. Lakes hundreds of feet deep filled the canyon. On his 1869 journey through here, John Wesley Powell observed the remnants of these eruptions and wrote: "What a conflict of water and fire there must have been here! ... What a seething and boiling of waters; what clouds of steam rolled into the heavens!" But the Colorado River, on its inexorable course to the sea, made short shrift of these natural dams. So, too, the river will eventually have its way with Glen Canyon Dam and will once again resume the task of sculpting its grandest work.

All these visions appear before the mind's eye as the hiker soaks burning feet in **Bright Angel Creek** where it enters the Colorado River. Boulders crunch and grind down the rollicking stream, at times sounding like voices as the water relates how it

made this canyon, how it opened the chapters of this epic book – of entire oceans, mountains, deserts, and rivers that have come and gone.

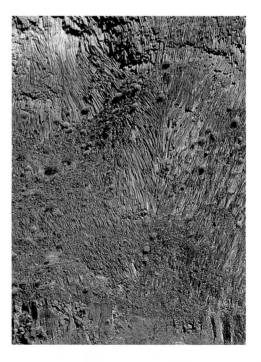

TRAVEL TIPS

DETAILS

When to Go

The South Rim is open year-round; the North Rim is closed from November to May. Spring and fall are pleasant. Summer is crowded, hot, dry, and prone to spectacular thunderstorms and sudden temperature drops. Winters are bitterly cold but marked by beautiful snowfalls and quiet.

How to Get There

Major airlines fly to Phoenix Sky Harbor International and McCarran International in Las Vegas. Commuter airlines serve Pulliam Field in Flagstaff and Grand Canyon National Airport in Tusayan. The Grand Canyon Railway (800-843-8724) runs a vintage steam engine 65 miles from Williams to the South Rim.

Getting Around

Free shuttle buses run from Tusayan to the South Rim and along the West Rim and East Rim Drives in summer. Shuttle service also connects the North and South Rims, but you'll want a car to explore the surrounding area. Car rentals are available at airports in Phoenix, Las Vegas, and Flagstaff.

Handicapped Access

The South Rim visitor center and some shuttle buses are accessible; write the park for a free guide to accessibility.

INFORMATION

Grand Canyon National Park

P.O. Box 129, Grand Canyon, AZ 86023; tel: 520-638-7888.

Grand Canyon Chamber of Commerce

P.O. Box 3007, Grand Canyon, AZ 86023; tel: 520-638-2901.

CAMPING

The South Rim has two campgrounds with more than 350 sites; the North Rim has about 80 campsites. There are about 60 backcountry campsites as well. To reserve a site, contact the National Park Reservation Service, P.O. Box 85705, San Diego, CA 92186-5705; tel: 800-365-2267.

LODGING

PRICE GUIDE – double occupancy

$ = up to $49 $$ = $50–$99
$$$ = $100–$149 $$$$ = $150+

All park lodging is coordinated by Grand Canyon National Park Lodges. For same-day reservations, call 520-638-2631; for advance reservations, call 303-297-2757.

Bright Angel Lodge

P.O. Box 699, Grand Canyon Village, AZ 86023; tel: 520-638-2631.

This rustic lodge, opened in 1935, offers hotel rooms and cabins on the edge of the canyon's South Rim. $$–$$$

El Tovar Hotel

P.O. Box 699, Grand Canyon Village, AZ 86023; tel: 520-638-2631.

Built in 1905 in the style of a European hunting lodge, this stone-and-log hotel offers some 78 rooms at the edge of the South Rim. The dining room is one of the finest in the area. Book lodging well in advance. $$$–$$$$

Grand Canyon Lodge

Grand Canyon National Park, P.O. Box 129, Grand Canyon, AZ 86023; tel: 303-297-2757.

Built in the 1920s, this limestone-and-timber lodge offers breathtaking views from the edge of the North Rim. The main structure has 40 rooms; cottages also are available. $$–$$$

Kachina and Thunderbird Lodges

P.O. Box 699, Grand Canyon Village, AZ 86023; tel: 520-638-2631.

These neighboring lodges offer basic motel accommodations on the South Rim. $$–$$$

Maswik Lodge

P.O. Box 699, Grand Canyon Village, AZ 86023; tel: 520-638-2631.

The Maswik is a two-story motel with more than 270 rooms and cabins, all with private baths. The lodge is open year-round, but the cabins are closed in winter. $$–$$$

Phantom Ranch

P.O. Box 699, Grand Canyon Village, AZ 86023; tel: 520-638-2631.

Set on Bright Angel Creek in the inner canyon, Phantom Ranch is the only lodging facility below the rim. Accommodations are men's and women's dormitories, cabins, and a campground. Accommodations and meals must be reserved well in advance. $–$$

Yavapai Lodge

P.O. Box 699, Grand Canyon Village, AZ 86023; tel: 520-638-2631.

One mile east of Grand Canyon Village near the South Rim visitor center, the Yavapai has 358 rooms in several modern, two-story buildings. The lodge is closed from December to March. $$–$$$

TOURS

Arizona Raft Adventures

4050 East Huntington Drive, Flagstaff, AZ 86004; tel: 520-526-8200.

Whitewater excursions on the Colorado River last six to 16 days.

Grand Canyon Dories

P.O. Box 216, Altaville, CA 95221; tel: 800-346-6277 or 209-736-0805.

Multiday excursions aboard dories specialize in side hikes to Matkatamiba Canyon, Deer Creek Falls, Elves Chasm, and Havasu Creek.

Canyon Explorations

P.O. Box 310, Flagstaff, AZ 86002; tel: 800-654-0723 or 520-774-4559.

Guided oar and paddle trips through the Grand Canyon include daily hikes up side canyons, interpretive talks, and camping. Trips last one to two weeks.

Grand Canyon Field Institute

P.O. Box 399, Grand Canyon, AZ 86023; tel: 520-638-2485.

Educational programs combine hiking and other travel in the canyon with examinations of canyon geology, archaeology, and history.

Grand Canyon Lodges–Amfac Parks and Resorts

14001 East Iliff, Aurora, CO 80014; tel: 303-297-2757 or 520-638-2631.

Mule rides, bus tours, and other Grand Canyon activities.

Outdoors Unlimited

6900 Townsend Winona Road, Flagstaff, AZ 86004; tel: 800-637-7238 or 520-526-4546.

The lower, upper, or entire Grand Canyon may be covered on these multiday trips. Passengers hike side canyons and take short excursions to Native American ruins and remote grottoes.

Excursions

Glen Canyon National Recreation Area

P.O. Box 1507, Page, AZ 86040; tel: 520-608-6404.
The completion of the Glen Canyon Dam on the Colorado River in 1963 created Lake Powell, flooding most of Glen Canyon itself. Most of the rock walls are beautifully eroded Navajo sandstone; sadly, many of the canyon's intricate geologic features were drowned by the lake. Explore the dam at Page, Arizona, take a boat trip to Rainbow Bridge (see below), and visit the mid-lake region at Bullfrog and Halls Crossing via Highways 95 and 276.

Paria Canyon–Vermilion Cliffs Wilderness

Bureau of Land Management, Arizona Strip Field Office, 345 East Riverside Drive, St. George, UT 84790; tel: 435-688-3230.
The Paria River flows south out of Grand Staircase-Escalante National Monument and into the Colorado River near Lee's Ferry. Along the way, it knifes through colorful layers of Triassic and Jurassic rocks. Experienced hikers can make challenging, multiday journeys into several slot canyons.

Much easier is a day hike on 1.7-mile Wire Pass Trail, which leads into a slot canyon from a trailhead on Highway 89 between Kanab, Utah, and Page, Arizona.

Rainbow Bridge National Monument

P.O. Box 1507, Page AZ 86040; tel: 520-608-6404.
At a height of 309 feet, Rainbow Bridge is the largest natural bridge in the world. Carved by Bridge Creek, a tributary of the Colorado River, the span is composed of 200-million-year-old Navajo sandstone, which stands on abutments of slightly older Kayenta sandstone. Travelers can reach the site on foot or horseback through Navajo tribal land (permit required) or get there by boat from marinas at Dangling Rope, Wahweap, Bullfrog, or Halls Crossing.

Carlsbad Caverns National Park

New Mexico

CHAPTER **15**

"**N**ow, remember the caver's rule: Stay in touch with the rest of the group at all times. Place your feet on the ladder, say clearly 'on ladder.' Then, at the bottom, don't forget to call 'off ladder,' so the next person can descend. Avoid touching or stepping on any cave formations – they are very fragile – but keep three points of contact on the surface to steady yourself. Okay, everyone. Hats and headlamps on. Let's go." ◆ Ranger Viv Sartori adjusts her hard hat and headlamp, and flashes a quick, reassuring smile at her slightly apprehensive charges before placing booted feet on the 30-foot steel ladder and disappearing into the darkness. One by one, the members of the group follow, calling their position to each other. ◆ Neophyte spelunkers one and all, they signed up for this wild cave tour as a first caving foray. After using a rope, descending and ascending ladders, tiptoeing through silent, decorated chambers, squeezing through a few tight spots, crawling along a corridor, and experiencing a blackout far down below, chances are they'll be hooked for life. ◆ This isn't just any cave.

Subterranean galleries adorned with glistening rock sculptures await visitors beneath the desert.

It's the fairly undeveloped **Lower Cave**, one of 80 known underground chambers at **Carlsbad Caverns**, one of the world's most remarkable cave systems. Here, in this southeastern corner of New Mexico, far below the Guadalupe Mountains, is a clammy, strange world no human could call home – a pitch-black province of dripping water, year-round 56°F temperatures, secretive migratory bats, specially adapted crickets and beetles, and millions of remarkable calcite formations. This is a place with its own logic, and a unique origin that has even the most jaded geologists jumping up and down with delight.

The Pearlsian Gulf in Lechuguilla Cave got its name from the thousands of "cave pearls" – small calcite balls – that litter the bottom of the underground lake.

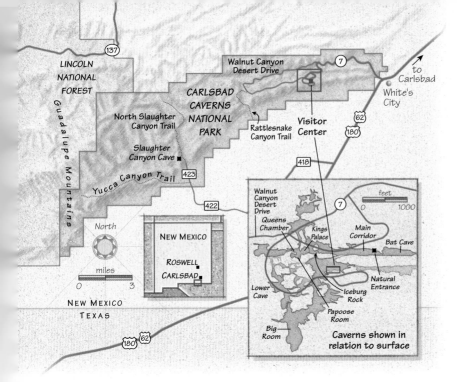

Caverns shown in relation to surface

potash, and other sedimentary deposits.

Then, between 70 million and 5 million years ago, a period of geological uplift gave rise to the Guadalupe Mountains and erosion began to strip away the sediments covering the fossil reef. The young mountains were particularly vulnerable to the hydrogen sulfide gas that seeped up into the water table from the oil reserves in the Permian Basin just to the southeast. The mixture of gas and the oxygen-rich water formed a mild sulfuric acid solution that dissolved large chambers at the level of the water table. As the water table dropped, caves formed at greater depths.

Underground Expeditions

New caves are being discovered all the time: The large, decorated chambers of **Slaughter Canyon Cave** (once called New Cave) were found in 1939. **Lechuguilla Cave**, explored in 1986 and open only for research purposes, has such strange and unearthly beauty that the National Geographic Society has mounted explorations there, NASA uses it as a stand-in for Mars, and most recently, scientists have detected cancer-killing microbes that may yet help win the war against this age-old plague.

But what draws visitors to

Pisoliths (above, left) are concretions of calcite that form around grains of sand.

A stalagmite (right) rises from the floor of the cave.

Stalactites (opposite) hang like icicles from a cavern ceiling.

Limestone with a Twist

Like many caves, Carlsbad Caverns formed in a thick bed of limestone, but with one significant difference. These caves are part of what was once a 400-mile-long, horseshoe-shaped reef growing on an offshore shelf of a shallow, inland sea in the Permian Period some 250 million years ago, when New Mexico lay close to the equator. The reef was composed of lime precipitated from seawater and limy secretions from sponges and calcareous algae. Eventually, this ancient sea became landlocked and evaporated, causing the reef to be buried under a heavy load of gypsum, salt,

Carlsbad Caverns National Park are the formations, or speleothems, that encrust its inner surfaces. It's impossible to date the formations (and attempts to do so in Lower Cave caused too much damage to warrant a repeat experiment), but cave scientists believe they began forming around half a million years ago, perhaps when the climate was wetter and cooler. Groundwater moving down cracks in the rocks encountered drier air in the caves, dropped its carbon dioxide load, and evaporated. Left behind were crystalline formations of calcite that look like they wouldn't be out of place in a Russian winter ice palace. Their regal appearance obviously struck others, too, who dubbed some of the chambers King's Palace, Queen's Chamber, Bifrost Room, and Papoose Room.

In the Hall of Giants, formations like Twin Domes are actually enormous stalagmites growing up from the floor. In other places, stalactites hang from the ceiling in great numbers and, where stalactites and stalagmites meet, great columns have grown up. Smaller iciclelike stalactites called soda straws hang in curtains, while eccentric helictites seem to grow in all directions. Smooth flowstone covers many sloping surfaces with marblelike deposits and, in places, slow-moving water has caused rimstone dams to form. Water is a constant here, forming pools and lakes that may eventually become obscured by calcite lily pads. Rarely, one may see delicate, needlelike aragonite formations, a product of similar depositional processes but with a different crystalline structure.

Come On Down

Carlsbad offers a full roster of self-guided and ranger-guided tours, from easy to technical. But all visitors start with a self-guided tour of the **Big Room**, one of the world's largest chambers and the centerpiece of Carlsbad Caverns. An elevator leads to the chamber, which offers easy viewing along a paved trail, an underground concession, seating for ranger talks, even restrooms. The more adventurous will want to enter the caves via the steep, one-mile paved trail leading from

A Sea of Sand

It comes into view like a bizarre desert mirage, 15 miles southwest of **Alamogordo**. White Sands National Monument is 275 square miles of ever-changing dunes that undulate across the **Tularosa Basin**, New Mexico's Space Central, a fitting place for such an alien landscape.

The snow-white sand is composed of gypsum, a hydrous form of calcium sulfate that is washed into the basin from the **San Andres Mountains** to the west and the **Sacramento Mountains** to the east. The runoff is trapped in low-lying ephemeral lakes that evaporate in the sizzling heat, leaving behind gypsum crystals that are blasted apart by extreme temperatures and winds.

The gypsum was originally deposited at the bottom of an ancient sea that became landlocked and evaporated about 250 million years ago. The sedimentary layers later domed up, then, about 10 million years ago, the center collapsed, creating the two ranges that flank the basin today.

The dunes are always on the move, constantly reshaped by winds that funnel between the mountains. Careful observers can identify at least four distinct types (see diagram, right). At the eastern edge of the basin are crescent-shaped parabolic dunes anchored by hardy plants such as four-wing saltbush and soaptree yucca. Barchan dunes, also crescent shaped even though their noses point into the wind and their arms trail behind, measure up to 65 feet high and are found in the middle of the basin, as are transverse dunes, which form long undulating ridges like a string of waves at sea. The park road exits through a band of dome dunes, the monument's youngest and most active, migrating some 30 feet a year.

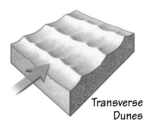

Parabolic
Dunes

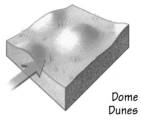

Transverse
Dunes

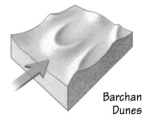

Dome
Dunes

Barchan
Dunes

the **Natural Entrance**. This lies past the **Bat Cave** passageway, summer breeding grounds for some 300,000 Mexican freetail bats, whose whirring exits at twilight and entrances at dawn are one of the park's biggest draws between May and October.

As you walk down, notice how quickly the shaft of light penetrating the gloom from the cave mouth, known as the Twilight Zone, recedes and an indefinable cave-time takes over. You can almost feel the cave breathing. Ears become accustomed to the incessant drip, drip of water, and eyes adjust to the low light and the phantasmagoric hordes of cave hoodoos and polymorphs hugging the trail.

Looking up, you see how the backreef area, with its characteristic mix of carbonates and evaporites laid down in a still lagoon, begins to give way to the reef itself, with its massive, vertically cracked limestone walls arching upward and glittering with gypsum.

At **Iceberg Rock**, pause and examine the rock face. Here, etched in bas-relief, are hundreds of tiny, ricelike fossils known as fusilinids. These and other tiny marine creatures such as trilobites, ammonites, nautoloids, and brachiopods were wiped out in a mass extinction at the end of the Permian, the greatest extinction in the fossil record.

Cave Commerce and Preservation

Considering the caves' great age, discovery of their treasures took a long time. Prehistoric Indians used the area around the cave mouth as a shelter and left behind painted symbols on the walls. Later, Mescalero Apaches gathering agaves for food left roasting pits. The Spanish, so eager for New World treasure, saw none of these natural riches – it was only when 19th-century cowboys discovered great deposits of bat guano that interest grew in exploiting the caves for commercial gain. Workers dropped down in buckets to harvest the natural fertilizer to sell to California's burgeoning citrus orchards.

One cowboy, Jim White, ventured farther in and was the first to report the beautifully decorated chambers. Few believed him until a photographer documented White's claims in 1915 and put the resulting photos on exhibit in Carlsbad. Tourists began paying for the privilege of being lowered in a bucket to the caves below. Jim White acted as guide and, when the caverns were set aside as a national monument in 1923, became chief ranger. Carlsbad gained the status of a national park in 1930.

Today, traveling into the caves is far easier but has brought its own problems. Pollution from the parking lot and maintenance yard enters the caverns through groundwater and damages the formations, most of which have stopped growing now that the climate is drier. Preservation of existing decorations has, therefore, become a priority. Much damage took place during the early years, when park staff and explorers thought nothing of giving away calcite "cave pearls," lighting fires that stained the rocks, and leaving handprints and writing on formations.

When you head into the caves now, make your watchword Leave No Trace. And if you want a little cave magic, Ranger Sartori has a tip for you. Take her Lower Cave tour, shut your eyes when she lets off her camera flash beside a column, then watch the formation glow. Its brief beauty is the message. Enjoy it while it lasts.

A caver (right) examines a "mushroom" formation in Slaughter Canyon Cave.

Yucca (opposite) is one of the few plants that can survive the extreme aridity of White Sands' gypsum dunes.

TRAVEL TIPS

DETAILS

When to Go

Underground temperatures at Carlsbad Caverns are a constant 56°F year-round. Bring a light jacket, rubber-soled shoes for traction on slippery trails, and gloves to protect formations on off-trail tours. The surrounding Chihuahuan Desert is most pleasant in fall and spring. Temperatures at low elevations exceed 100°F in summer; snow is likely in winter.

How to Get There

Major airlines serve El Paso, Texas, about 165 miles away, and Albuquerque, New Mexico, about 275 miles away. Mesa Airlines connects directly to Carlsbad. Amtrak stops at Albuquerque and Deming, New Mexico, and El Paso. Greyhound and TNM&O bus lines serve Carlsbad.

Getting Around

A car is the only convenient means of travel in the area. Rentals are available at the airports.

Handicapped Access

The visitor center, much of the self-guided Big Room Tour, an outdoor trail, and picnic sites near the visitor center and at Rattlesnake Springs are wheelchair-accessible. Contact the park's handicapped-access coordinator for detailed information.

INFORMATION

Carlsbad Caverns National Park

3225 National Parks Highway, Carlsbad, NM 88220; tel: 505-785-2232 (visitor information) or 800-967-2283 (reservations).

Carlsbad Convention and Visitors Bureau

302 South Canal Street, P.O. Box 910, Carlsbad, NM 88220; tel: 800-221-1224 or 505-887-6516.

New Mexico Department of Tourism

491 Old Santa Fe Trail, P.O. Box 20002, Santa Fe, NM 87503; tel: 800-733-6396.

CAMPING

Pine Springs Campground

Guadalupe Mountains National Park, HC 60, Box 400, Salt Flat, TX 79847; tel: 915-828-3251.

This campground, 32 miles south of Carlsbad, has 20 sites and fills up quickly in summer.

LODGING

PRICE GUIDE – double occupancy

$ = up to $49 $$ = $50–$99
$$$ = $100–$149 $$$$ = $150+

Best Western Cavern Guadalupe Inn

17 Carlsbad Highway, Whites City, NM 88268; tel: 800-228-3767 or 505-785-2291.

This 63-room motel has three sections: Walnut Inn, Cavern Inn, and Guadalupe Inn. Guadalupe Inn is built around a courtyard and offers spacious rooms with ceiling beams and Southwestern furnishings. A small waterpark is a new addition at Cavern Inn. A large swimming pool, saloon, and restaurant are on the premises, along with an RV park with restrooms, showers, laundry facilities, and pool privileges. $$

Best Western Motel Stevens

1829 South Canal Street, Carlsbad, NM 88220; tel: 800-730-2851 or 505-887-2851.

This chain motel has 204 spacious rooms, some with microwaves and refrigerators. A pool and restaurant are on the premises. Shuttle service is provided from the airport. $$

Carlsbad Holiday Inn

601 South Canal Street, P.O. Box 128, Carlsbad, NM 88220; tel: 800-465-4329 or 505-885-8500.

Set in downtown Carlsbad about 23 miles from the caverns, the hotel offers 100 rooms furnished in Southwestern style, some with whirlpool baths. A restaurant specializes in gourmet dining; casual fare is available in an adjoining bar and grill. A mist-cooled patio, swimming pool, sauna, whirlpool, workout room, and laundry facility are on the premises. Room service is available. $$

Continental Inn

3820 National Parks Highway, Carlsbad, NM 88220; tel: 505-887-0341.

A good choice for those traveling on a budget, this two-story red-brick motel has 60 modest rooms with oak furniture and coffee makers. Suites have large sleeper sofas and can accommodate small groups. A swimming pool is on the grounds. $

Horseshoe Hacienda

84 Means Road, Carlsbad, NM 88220; tel: 800-607-4600 or 505-785-2213.

This bed-and-breakfast, four miles from Carlsbad Caverns and 40 miles from Guadalupe Mountains National Park, has seven guest rooms, four with private baths. Hiking and horseback riding are available. Inquire about special chuck-wagon dinners and wagon rides. $$

TOURS

Carlsbad Caverns National Park

3225 National Parks Highway, Carlsbad, NM 88220; tel: 505-785-2232 (visitor information) or 800-967-2283 (reservations).

The Park Service operates all tours of Carlsbad Caverns. Admission to the Big Room via the elevator or the one-mile Natural Entrance route is included in the entrance fee. Visitors take a self-guided tour or one of several daily ranger-led tours. Tours of additional rooms, led by rangers, must be reserved in advance. Off-trail tours of varying difficulty, such as Left Hand Tunnel and Lower Cave, are restricted to 12 participants; reservations are required. All tours depart from the visitor center, except the tour of Slaughter Canyon Cave, which is reached via an 11-mile side road from Highway 180.

MUSEUMS

The Space Center

Scenic Drive and Indian Wells Road, Alamogordo, NM 88311; tel: 800-545-4021 or 505-437-2840.

Visitors examine Moon rocks and other aspects of extraterrestrial geology at this major air and space museum.

Waste Isolation Pilot Project Facility

Carlsbad, NM, tel: 800-336-9477.

Tour the nation's first deep-geologic disposal facility for defense-generated radioactive waste. Call for reservations at least two months in advance.

Excursions

Guadalupe Mountains National Park

HC 60, Box 400, Salt Flat, TX 79847-9400; tel: 915-828-3251.

The same limestone reef that contains Carlsbad Caverns has been uplifted and eroded into crags and canyons in nearby Guadalupe Mountains National Park, site of Texas' highest mountain, 8,749-foot Guadalupe Peak. More than 80 miles of trails crisscross the desert lowlands and lead into the high country. Most popular is McKittrick Canyon, where Texas madrone, oak, and other deciduous trees put on a brilliant display each fall.

Big Bend National Park

Big Bend National Park, TX 79834; tel: 915-477-2251.

Encompassing 1,200 square miles on the Rio Grande, Big Bend preserves a sprawling chunk of the Chihuahuan Desert, three spectacular river canyons (Santa Elena, Mariscal, and Boquillas), and about 5,000 plant and animal species. In the center of the park are the Chisos Mountains, formed principally by volcanic activity some 35 million years ago. The Sierra del Carmen, a much older range to the east, was uplifted with the rest of the Rocky Mountains about 75 million years ago and is composed of limestone deposited at the bottom of ancient seas.

Chiricahua National Monument

Dos Cabezas Route, Box 6500, Willcox, AZ 85643; tel: 520-824-3560.

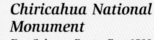

The bizarre spires, pinnacles, pedestals, and balanced rocks at the heart of southeastern Arizona's Chiricahua Mountains are an unexpected sight amid the verdant highlands that attract both Mexican and American wildlife. The gray rhyolite hoodoos owe their existence to an eruption of white ash from nearby Turkey Creek caldera 25 million years ago. This material hardened, uplifted, and eroded into a rogue's gallery of geological forms that are best viewed on trails from Massai Point.

Death Valley National Park

California

CHAPTER **16**

aught in a rain shadow between the Rockies in the east and the Sierra Nevada in the west, the **Great Basin** is aptly named. With no outlet to the sea, rivers flow in – and then evaporate into thin, and very hot, air. The salt flats left behind may be useful for salt, borax, and gypsum mining, but are hardly life-sustaining. ◆ The Basin, a 250,000-square-mile desert covering western Utah, eastern California and Oregon, and most of Nevada, contains hundreds of mountain ranges running north to south, bounded by faults and small basins. One geologist likened them to a fleet of ships at anchor pointing into a north wind. ◆ **Death Valley National Park** in eastern California is the Great Basin at its most mysterious. The park covers 3.3 million acres, the largest national park in the Lower 48. It once made the record books as the hottest place on Earth (134°F recorded July 10, 1913, at **Furnace Creek**) and remains one of the driest (an annual rainfall of about 1.9 inches, combined with an evaporation rate of 120 to 150 inches a year).

The Earth bares its weathered bones at this heat-blasted park in the California desert.

Record heat and aridity feature in all descriptions, but few visitors expect so many high mountains, cool side canyons and springs, not to mention the more than three million acres of wilderness protected in the park. ◆ Death Valley was named by a party of forty-niners who nearly died here, but actually it's filled with life – 440 vertebrates and nearly 1,000 plant species, including 3,000-year-old bristlecone pines. It rivals the Grand Canyon as a place to see both extremely old and extremely young rocks – sometimes right next to each other. And with its jumbled geology, it remains a headscratcher for earth scientists.

A lone hiker walks along a ridge in Eureka Dunes; such dunes are rare in Death Valley because salt cements most loose grains together.

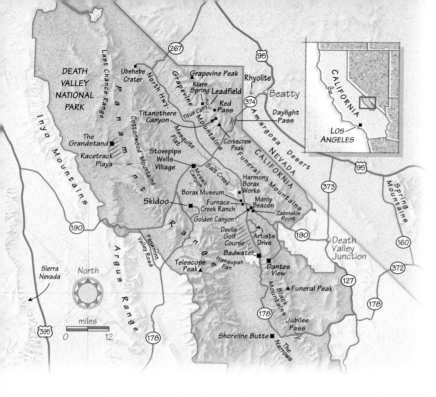

for a panorama of Death Valley and its surroundings. Aim for early morning, when the air is suffused with pastels, and shadows throw formations into dramatic relief. Stepping to the north, a thousand feet higher, are the **Funeral Mountains**, and beyond, the **Grapevines**, topped by 8,738-foot **Grapevine Peak**. To the east lie the **Amargosa Desert** and several Great Basin ranges, including the **Spring Mountains** topped by 11,918-foot **Mount Charleston**, near Las Vegas. Fifteen miles west of Dante's View is 11,049-foot **Telescope Peak** and the **Panamints**, while beyond are the **Argus** and the **Inyos**. On a clear day you can see the Sierra Nevada and 14,000-foot peaks like Mount Williamson.

Although its origin is imperfectly understood, the Great Basin's topography is probably connected to the disappearance of the ancient Farallon tectonic plate beneath the North American plate's western edge. This event created the San Andreas Fault and led to an extension of the Earth's crust, which is much thinner in the Great Basin and therefore a good deal hotter. The land under the Basin is literally being pulled apart – an estimated 150 miles to date and growing. California's famous restlessness may be more than a state of mind after all.

Land of Extremes

The first stop in the park should be 5,475-foot **Dante's View** in the **Black Mountains**

But what most people remember is the Death Valley basin, which opened up between the Panamint and Black Mountains about three million years ago. Rivers in flood, full of boulders, gravel, silt, and sand, have cut canyon narrows in the mountains and poured into the basin, dropping their loads as huge alluvial fans. Directly across from Dante's View is the **Hanaupah Fan**, one of the oldest in the park. Its loose debris has hardened over time into a beige "fanglomerate" rock, as geologists call it, and has been delicately

Death Valley geology (opposite) is a gallery of textures and patterns. Here, from left to right, are conglomerate rock, mud cracks, and mineral patterns.

Salt crystals (right) growing upward in mudcracks have an eerie resemblance to a sheet of pack ice. The Amargosa Range stands in the background.

etched by creeks and faulting.

In the central basin, white salt flats shimmer like mirages, the remnants of Pleistocene glacial lakes that once filled the valley. These evaporated, leaving behind ponds, salt flats, and briny streams like **Salt Creek**, conducive only to pickleweed, salt grass, and relict pupfish. Today, water draining from surrounding mountains continues the cycle, adding to the evaporites in the valley.

The lowest point in the Western Hemisphere sits across from Telescope Peak, near **Badwater**, at 282 feet below sea level. The two extremes constitute the greatest elevation change in the least distance in the continental United States.

Imaginary Lake

So what's going on here? A tour of some of Death Valley's hot spots (and be warned, temperatures regularly reach 120°F in summer) offers some clues. Fill your gas tank and buy plenty of water and supplies at **Furnace Creek Ranch** off Highway 190, stop for information at the visitor center, then drive south on the paved scenic road to Badwater. Park the car and look back at the mountains.

The Black Mountains began to rise some 14 million years ago, an eyeblink in geologic time, jounced up along the Death Valley Fault Zone. This fault zone remains active.

The floor of Death Valley has been subsiding, sliding down the west face of the Black Mountains. As the bedrock drops, sediments carried from surrounding highlands build up in the basin. Once deposited, the sediments are themselves tipped as the floor continues to rotate downward. Today, there are two miles of sediments below the surface of Badwater.

For the quintessential Death Valley experience, walk out onto the salt pan at Badwater and imagine the seas and lakes that once lapped in what is now basin and range. You

can see evidence of Pleistocene Lake Manly – actually a series of lakes once 600 feet deep – in the wave-cut terraces in **Shoreline Butte** and the gravels beside **Jubilee Pass** in south Death Valley, as well as in the valley's mud hills and salt flats. The outside of the salt flats contains sand and silt deposits covered in calcium carbonate, which shrink up into mounds as they dry, leaving the kind of moonscape you see at **Devils Golf Course**. Closer in are sulfate and gypsum deposits. Then, at the center is sodium chloride – lowly table salt.

The Panamints contain granites and other igneous rocks, as well as a large section of Paleozoic sedimentary rocks, 540 to 245 million years old. But the Blacks contain both ancient Precambrian rocks and young volcanic deposits in an interesting juxtaposition. The oldest rocks were laid down in the ocean 1.7 billion years ago, when the West Coast shoreline lay near Las Vegas, and were later metamorphosed into gneiss. The youngest rocks were deposited as sediments and volcanic ash and basalts in the recent Cenozoic era.

The Colors of Youth

For a look at younger rocks, take an afternoon spin through **Artist's Drive**, near Badwater. The Artist's Drive Formation, awash in colorful manganese – and iron-rich yellows, purples, greens, pinks, and browns – is between 14 and 7 million years old and consists of muds and volcanic ash from Tertiary-era eruptions that blanketed the region. **Golden Canyon**, which

Manly Beacon (top) rises above easily eroded badlands.

The view from Zabriskie Point (left) takes in the surrounding mud hills and the distant Amargosa Mountains.

Sliding Rocks

One of Death Valley's most compelling mysteries concerns the sliding rocks of **Racetrack Playa** in the park's remote northwest corner. What exactly causes boulders – some weighing as much as 700 pounds – to sail across the dry lakebed, leaving tracks in dried mud as evidence of their passage? No one knows for sure.

Named for its oval shape and the dark formation known as the **Grandstand** on its north side, the Racetrack lies between the **Cottonwood Mountains** and the southern section of the **Last Chance Mountains**. It is reachable only by a long, rugged road. Muddy sediments wash down from the Cottonwoods onto the basin floor, where they eventually dry and crack into polygon shapes. The wind fills the cracks with sand.

When wet, usually in winter, the flat surface of the Racetrack is covered in fine wet clay, slick enough perhaps for rocks to slide on during high winds. Some researchers suggest an additional factor: ice floes around the rocks. But with rocks moving only every two or three years and tracks that last not much longer, who knows when an answer will be found? And to date, not a soul has ever witnessed the rocks moving.

The Racetrack's sliding rocks (left) remain a mystery; no one is quite sure how they move across the desert floor.

Wagons (below) pulled by teams of 20 mules were used to haul borax out of Death Valley.

another, causing highly polished surfaces, or "slickensides."

But geologists offer differing explanations for their origin. One intriguing theory has it that the Panamints slid off the Blacks about six million years ago to become a separate range, leaving behind this exposed detachment zone. Another theory holds that the Panamints may have once sat beside the Blacks, then were pulled to their present position by crustal extension. The mountain blocks may have rotated as they slipped down along curved faults, creating the tilted mountains and asymmetric basins.

offers one of the best valley hikes, contains hematite-rich rock that ancestors of the Timbisha Shoshone Indians, who still live in and around the park, used for red paint. **Zabriskie Point**, with its badlands of hardened mud hills and landmark **Manly Beacon**, is another place to see younger deposits.

In fact, young hills of crumbly sediments line much of the east side of the park and were the source of Death Valley's successful borax mines in the late 1800s. To learn more about the mining operations and famous Twenty Mule Team wagons that transported the borax, visit **Harmony Borax Works** and the **Borax Museum** near Furnace Creek.

Colorful, convoluted, ledgy rocks with a pronounced northwest tilt are a feature of these ranges. Some of the strangest can be seen bordering the highway at Badwater, Mormon Point, and Copper Canyon. Geologists have dubbed them turtlebacks and they remain one of the park's greatest mysteries. It is believed that they are exposed fault surfaces where rocks have rubbed against one

Flowing water cut a channel in Mosaic Canyon (right) during a wetter period in the valley's history; little erosion happens now.

The Amargosas (below), like most of the mountains around Death Valley, are tilted fault blocks whose roots are buried by alluvial fans of debris.

Boom and Bust

In the 19th century, folded and colored rocks were clear evidence to prospectors like Shorty Harris that gold, silver, lead, and other precious minerals lay "in them thar hills." Their practiced eyes sought out quartzite and granitic intrusions, such as those seen in the 100-million-year-old granites of **Skidoo** in the Panamints, Death Valley's most successful gold mine. Shorty cared little for money but loved the thrill of the chase for gold. To view another of his claims, head northeast out of the park over **Daylight Pass**, which follows the Boundary Fault between the Grapevine and Funeral Mountains. To the north is **Corkscrew Peak**, the axis of a large fold that has turned the rocks upside down, exposing the very red, 600-million-year-old Carrara Formation.

Shorty found gold near **Rhyolite** but sold his Bullfrog claim for a handful of cash and moved on. Today, you can tour the well-preserved ghost town, with its famous Bottle House, railroad station, and other buildings, all within sight of the Bullfrog mine and its ponds (now closed) near **Beatty**, Nevada.

The Beatty road is also the access point for Death Valley's best backcountry drive: 27-mile-long **Titus Canyon**. Use a high-clearance, four-wheel-drive vehicle to negotiate this one-way route. The narrow, winding, rough road across the southern Grapevine Mountains begins in the 3,000-foot **Amargosa Valley**, then heads north across the crest of the Grapevines among iron-rich riverbed sediments at **Red Pass**. A side hike takes you into **Titanothere Canyon**, where two rhinoceroslike titanotheres from 30 million years ago were uncovered in 1933. The road drops down to the 1926 lead mining community of **Leadfield**, one of Death Valley's most outrageous boondoggles – so far from civilization and marginal that it quickly failed and was abandoned.

Ubehebe Crater (right) at the northern end of Death Valley is the product of a relatively recent volcanic explosion.

Mushroom Rock (below), a large chunk of basalt, has had its lower section eroded away by wind-blown particles.

You then enter upper Titus Canyon, through beautiful limestone cliffs. Ancient Indian petroglyphs cover rocks in the oasis around **Klare Springs**. Bring your binoculars. The springs are a favored watering hole for shy bighorn sheep that are sometimes glimpsed by lucky visitors. Down-canyon, the road enters an anticline where you're treated to remarkable exposures of gray Cambrian Bonanza Formation, which closes in to less than 20 feet, before the road exits down a large alluvial fan to the highway. You can also drive up to the canyon entrance from Death Valley, park, and hike into the **Narrows**.

Dunes and Craters

Not far from Titus Canyon is **Stovepipe Wells**, the other main developed area in the park. It sits beneath **Tucki Mountain** and **Mosaic Canyon**, a wonderful place to take a shaded hike among upturned rocks – breccia, and Precambrian deposits that have been smoothed by floodwaters. Just north of Stovepipe Wells is **Mesquite Dunes**, the most accessible and photographed of Death Valley's five dune fields. The sands were picked up by north–south winds and blown through the valley, then dropped as the wind slowed near the mountains. This dune complex is unique because it's one of the only places in the world where you can see three distinct dune types – longitudinal, transverse, and star – in close proximity.

Northwest of here, you'll find clear proof that volcanic forces continue to rock Death Valley from time to time. Half-mile-wide **Ubehebe Crater** was born violently just 3,000 years ago. When intruding magma met groundwater, the water flashed into steam and exploded, leaving behind this 500-foot-deep crater and several sibling holes nearby.

"This is the tectonic, active, spreading, mountain-building world," commented one geologist. Standing on the wind-blown rim of this black crater, it is easy to believe.

TRAVEL TIPS

DETAILS

When To Go

The most comfortable time to explore Death Valley is November to April. Winter brings scant precipitation, which, when it does come, ensures spectacular wildflower displays in spring. Summer temperatures average well above 100°F and often reach a dangerous 120°F.

How to Get There

Major airlines, Amtrak, and interstate buses serve Las Vegas, Nevada, about two hours northeast of Death Valley.

Getting Around

A car is essential to get around in this enormous park; rentals are available at the airport. High-clearance, four-wheel-drive vehicles are necessary on unpaved, backcountry roads. Fuel is available at Furnace Creek, Scotty's Castle, Panamint Springs, and Stovepipe Wells within the park; gas stations are few and far between elsewhere.

Handicapped Access

Furnace Creek Visitor Center is handicapped-accessible, as are paved roads that access viewpoints and overlooks, and Salt Creek Nature Trail, a boardwalk suitable for wheelchairs. All other trails are rough routes. Call the park's handicapped-access coordinator for details.

INFORMATION

Death Valley Chamber of Commerce

P.O. Box 157, Shoshone, CA 92384; tel: 760-852-4524.

Death Valley National Park

P.O. Box 579, Death Valley, CA 92328; tel: 760-786-2331.

Mojave Desert Information Center

72157 Baker Boulevard, Baker, CA 92309; tel: 760-733-4040.

Needles Information Center

707 West Broadway, Needles, CA 92363; tel: 760-326-6322.

CAMPING

Death Valley National Park

P.O. Box 579, Death Valley, CA 92328; tel: 760-786-2331.

There are nine campgrounds in the park, totaling more than 1,500 individual sites. Several campgrounds, including Sunset, Furnace Creek, and Stovepipe Wells, have water and toilets. Some sites may be reserved; call 800-365-2267. Fall and early winter are the busiest seasons. Backpackers can camp at least two miles from developed areas, paved roads, or day-use-only areas; a permit is required for backcountry travel.

LODGING

PRICE GUIDE – double occupancy

$ = up to $49 $$ = $50–$99
$$$ = $100–$149 $$$$ = $150+

Amargosa Hotel

Death Valley Junction, CA 93522; tel: 760-852-4441.

Built in 1924, this adobe hotel, about 27 miles from the park, retains its Old West atmosphere. There are 10 guest rooms with private baths, an ice cream parlor, and dance shows in the adjoining Amargosa Opera House. $–$$

Furnace Creek Inn

AmFac Resorts, P.O. Box 1, Highway 190, Death Valley National Park, CA 92328; tel: 800-236-7916 or 760-786-2345.

This elaborate complex in the heart of Death Valley has landscaped grounds, hacienda-style buildings with red-tiled roofs, and 66 deluxe rooms. Amenities include tennis courts, a pool fed by hot springs, and a formal dining room. $$$$

Furnace Creek Ranch

AmFac Resorts, P.O. Box 1, Highway 190, Death Valley National Park, CA 92328; tel: 800-236-7916 or 760-786-2345.

Set amid date palms near the visitor center, the ranch offers about 220 simple, air-conditioned cottages. A swimming pool, golf course, general store, coffee shop, and steakhouse are on the premises. $$–$$$

Panamint Springs Resort

P.O. Box 395, Ridgecrest, CA 93556-0395; tel: 775-482-7680.

This rustic lodge is about an hour east of Furnace Creek and offers budget accommodations. Rooms have queen- and king-sized beds, armoires, and private bathrooms. $$

Stovepipe Wells Village

Death Valley, CA 92328; tel: 760-786-2387.

Stovepipe Wells Hotel, built in 1926, has 83 modest, air-conditioned rooms as well as a saloon, swimming pool, general store, gift shop, and restaurant. The Park Service operates 14 RV hookup sites on the premises. $$

TOURS

ATV Action Tours

Black Mountain Business Park, 175 Cassia Way, Suite A118, Henderson, NV 89014-6643; tel: 702-566-7400 or 888-288-5200.

This outfit leads small groups on excursions in and around Death Valley; tours involve light hiking.

Grand Canyon Tour Company

4894 Lone Mountain Road, PMB 137, Las Vegas, NV 89130; tel: 702-655-6060 or 888-512-0075.

Bus tours of Death Valley and other desert locations depart from Las Vegas daily.

Rocky Trails, Inc.

P.O. Box 371324, Las Vegas, NV 89137-1324; tel: 702-869-9991.

Trained geologists and naturalists lead nature tours of Death Valley in vans and four-wheel-drive vehicles; some hiking is required.

MUSEUMS

Harmony Borax Museum

Furnace Creek Ranch, Death Valley, CA 92328; tel: 760-786-2345.

Exhibits at this park museum chronicle Death Valley's short-lived borax-mining industry. The museum is housed in an 1883 wooden building, the oldest in the valley.

Maturango Museum

100 East Las Flores Avenue, Ridgecrest, CA 93555; tel: 760-375-6900.

The museum is dedicated to the natural and cultural history of the Mojave Desert. Visitors can participate in lectures, wildflower walks, and tours of Indian petroglyphs. The Death Valley Tourist Center and West Mojave Visitor Center share the site.

Shoshone Museum

P.O. Box 38, Shoshone, CA 92384; tel: 760-852-4524.

The collection specializes in geology, mining, and other aspects of desert life.

Excursions

Great Basin National Park

Highway 488, Baker, NV 89311; tel: 775-234-7331.

The jewel of this desert park, 100 miles west of Baker, is Lehman Caves, a huge marble-and-limestone cavern discovered by a rancher in the late 1800s. Stalactites, stalagmites, helictites, soda straws, and popcorn abound, along with an unusual formation called a shield, which resembles a clamshell. On the surface is 13,063-foot Wheeler Peak, the second highest in Nevada, and a remnant of an Ice Age glacier.

Joshua Tree National Park

74485 National Park Drive, Twentynine Palms, CA 92277; tel: 760-367-5500.

This 800,000-acre park is a magnet for rock climbers, for a good geological reason: The hornblende crystals found in the huge granite boulders grip sneakers and don't let bare skin slide. Wonderland of Rocks, a 12-square-mile granite maze, hides groves of Joshua trees, oases, and dry washes; the Jumbo Rocks area is also well-known for extraordinary rock formations and first-rate climbing routes. The rough, 18-mile Geology Road Tour just west of Jumbo Rocks leads through a landscape of dry lakebeds and giant boulders.

Mojave National Preserve

P.O. Box 241, Baker, CA 92309; tel: 760-733-4040.

Congress created this 1.6-million-acre preserve in 1994 to protect habitat for desert tortoises and other Mojave species within California's Lonesome Triangle, the area between Interstates 15 and 40. Don't miss 45-square-mile Kelso Dunes, the most extensive dune field in the West, and Cima Dome, a geologic anomaly topped by Joshua trees. A two-mile trail takes you among these giant, shaggy members of the lily family, connecting Cima Dome and Hole-in-the-Wall Campgrounds. The preserve is undeveloped, with no services or lodgings.

Yosemite
National Park
California

Yosemite occupies mythical as well as physical space. It's not a mere place but a kind of Platonic ideal, a venue that looks and feels magical simply because the scenery is so awesome: the flower-strewn meadows and deep woods of fir and pine, the rushing water, the merciless light and deep-blue sky, and above all, the stone, the gleaming white-and-gray granite. ◆ Like the face of a beautiful woman, Yosemite has excellent bones. They're the first thing you notice about the place: the great superstructures of rock buttressing the dales and rills and forests, reaching thousands of feet into the sky as soaring walls, spires, and pinnacles. ◆ But Yosemite didn't start out as a Shangri-la of granite formations. Indeed, its oldest rocks are limestone, chert, shale, and sandstone, sediments laid down during the Paleozoic Era 500 million years ago. During this time, Yosemite was virtually beachside property, the western margin of what is now North America. As rocks to the east eroded, thick

Towering cliffs, carved by ice and water, lie at the granite heart of the Sierra.

layers of sediment washed into the sea that lapped the continent. This sand, silt, and gravel combined with calcium carbonate and silica from marine life to produce the outcrops of sedimentary rocks found today in the park. ◆ During the Mesozoic Era, the Yosemite area underwent great upheaval, the result of the North American plate overriding the adjacent oceanic plate. This was the beginning of the region's igneous period, when Earth's interior fires forged Yosemite's signature granite. Considerable volcanism also occurred at this time, evidenced by masses of porphyry and ancient ash deposits in and around the park.

Granite and glaciers are the geological themes of Yosemite; both helped to create Bridalveil Falls, the Cathedral Rocks, and Sentinel Rock.

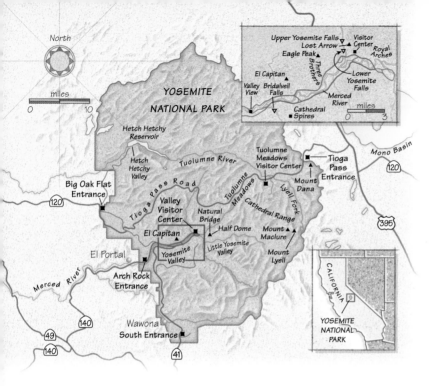

North

miles
0   10

YOSEMITE
NATIONAL PARK

Hetch Hetchy
Reservoir

Hetch
Hetchy
Valley

Tuolumne River

Big Oak Flat
Entrance
120

Valley
Visitor
Center

Natural
Bridge

El Capitan

Little Yosemite
Valley

El Portal

Yosemite
Valley

Arch Rock
Entrance

Merced River

140

49
140

Wawona
South Entrance

41

Tuolumne
Meadows
Visitor Center

Mount
Dana

Half Dome

Mount
Maclure

Mount
Lyell

Tioga
Pass
Entrance

120

Mono Basin

395

Tioga Pass Road

Tioga

Tuolumne Meadows

Lyell Fork

Cathedral Range

Upper Yosemite Falls
Lost Arrow
Eagle Peak

Visitor
Center
Royal
Arches

El Capitan

Valley
View

Bridalveil
Falls

Three Brothers

Lower
Yosemite
Falls

Merced
River

Cathedral
Spires

miles
0   3

CALIFORNIA

YOSEMITE
NATIONAL
PARK

Hot Heart of the Sierra

The granite that now underlies much of the central Sierra Nevada was formed during the Mesozoic between 225 million and 85 million years ago. Geologists call granite a plutonic igneous rock, meaning that it originates deep within the Earth. It comes from magma, or molten proto-rock.

The chemistry and thermodynamics of molten rock are both delicate and marvelous. If magma creeps to the surface at temperatures of roughly 1,000°C, the resulting rock (when cool) will be essentially granitic. But if the magma is hotter, about 1,200°C, it will be basaltic. Basalt is a sooty, monochromatic – and, frankly, unlovelier – rock than granite, which is speckled like a trout, with dark bits of hornblende and biotite

sprinkled through a light background of quartz and feldspar.

Yosemite's granite is part of a gargantuan balloon of rock known as the Sierra Nevada batholith. This tremendous granitic mass forms the foundation for much of the central Sierra, but it did not spring from the Earth's mantle in a single cataclysmic event. Rather, it represents numerous periods of intrusion, accretion, and solidification that produced a complex mélange of granitic and quasi-granitic rocks. Geologists have mapped about 100 separate masses of plutonic rock in the immediate vicinity of the park alone.

So what was the result of all this plutonic activity? Not today's Sierra Nevada but an impressive mountain range nevertheless. It was a long line of ridges and peaks parallel to the continent's seaboard – a kind of ancestral Sierra. According to N. King Huber, a research geologist with the U.S. Geological Survey who has studied the central Sierra, some of

Half Dome (right), formed from a single knob of granite, stands 4,700 feet above the valley floor.

the peaks were nearly 14,000 feet high.

By the late Mesozoic, things had quieted down in Yosemite. Magma activity shifted eastward, effectively ending the production of batholiths. Volcanism also generally ceased. The chief force working on the ancestral Sierra during this geologically slack era was erosion – water scraping away the spanking new mountains and dumping the debris in what is now California's Central Valley. In some places, the deposits are tens of thousands of feet thick. By the middle Cenozoic Era, the range resembled the Catskills more than the current Sierra, with elevations of a few thousand feet.

But things started to change during the late Cenozoic, about 20 million years ago. The oceanic plate that was pushing under the ancestral Sierra was totally consumed, and a northwest-moving

plate called the Pacific Plate took its place. The zone where these two plates met is today's San Andreas fault. Stresses built, and the Sierras began rising again, tilting up and westward, with an abrupt escarpment forming on the east and a long slope dropping on the west.

At the same time, water and volcanism again came into play. Streams coursed down the new cordillera, particularly on the long westward slope. Volcanoes bloomed at the northern end of the range, merging with the volcanic peaks of the Cascades. Just north of Yosemite National Park, the terrain was blanketed by volcanic ash, lava, and mudflows.

A granite boulder (top) rests near an oak; Yosemite's granite formed deep in the heart of an earlier mountain range and was brought to the surface by tectonic faulting and erosion.

The valley's spires and pinnacles (right) have drawn climbers for more than a century.

Restless Rock

Probably the most salient elements in Yosemite's current granitic landscape are due to joints – the geologic term for large-scale cracks in rock. Yosemite granite is rich in joints, and these largely determined the shape of some of the park's best-known features. Granite is hard and particularly resistant to erosion. If the batholith were smooth as an egg, it would take eons before any notable features would appear on its surface. But joints allow water and air into the heart of the stone. This weakens large masses of rock and lets gravity go to work. As joints deepen, large portions of rock slough off, creating newer, sheerer, more vertical structures.

Joints form while the granite is still crystallizing under great pressure deep in the Earth. When granite is uplifted and overlying rocks erode, the pressure all but vanishes, causing

the granite to expand. This creates new stresses, ultimately popping off the outer layers of rock in a process called exfoliation, to which Yosemite's granite is particularly prone. Over time, exfoliation rounds irregular granite masses into domes and walls that exhibit peeling slabs of rock like layers of an onion. The **Royal Arches**, a massive series of half circles broken out of the rock east of El Capitan, are prime examples of exfoliating rock.

But while joints shape Yosemite's landscape, they are not the whole story. Certain portions of the batholith formed relatively free of joints. These have evolved into some of the park's most dramatic features, notably **El Capitan** and **Half Dome**. These are massive simply because their rock is comparatively unflawed and most of it remains intact.

When unfractured and unexposed to moisture, granite is virtually inert. But chemical reactions occur when it combines with water, soil, and

El Capitan (opposite), a single, relatively unflawed piece of granite that rises more than 3,000 feet from valley floor to summit, is not impervious to erosion. Thin slabs of granite tend to peel off the rock face (top) and fall away.

Yosemite Falls (left) drops 2,500 feet from a hanging valley in two cascades separated by a rocky chute.

organic compounds, causing it to decay quite readily. Sometimes water scours away the softened granite, leaving harder rock behind. **Natural Bridge** on Half Dome is a dramatic example of this process.

Water has also shaped the park on much larger scales. The **Tuolumne** and **Merced Rivers** exemplify the role streams play in carving landscapes. As the Sierra lifted during the late Cenozoic, streams and rivers grew larger and swifter due to the steeper gradient. The mountains rose faster than erosion could scour away their upper elevations. In such situations, waterways tend to cut deep canyons, and that's exactly what happened with the Tuolumne and Merced drainages. Today, these two gorges are among the Sierra's deepest.

The Sculpting Hand of Ice

Ice has also changed Yosemite. Though glaciers probably helped shape the Sierra long before, only the last two periods of glacial activity – the Pre-Tahoe and Tioga glaciations – can be clearly discerned. And of the two periods, the Tioga is responsible for most of the park's visibly glaciated features.

The Tioga glaciation began 30,000 to 60,000 years ago, when a cooling climate let small glaciers form high in the Sierra. As the chill deepened, these incipient glaciers grew and ultimately converged into vast icefields, with only the range's highest peaks remaining uncovered. Between 15,000 and 20,000 years ago, the icepack reached maximum size. At the time, ice clogged the Tuolumne River from its upper basin to the **Hetch Hetchy Valley**, and the Merced River through **Little Yosemite Valley**. On the South Fork of the Merced, the ice almost reached the site where the Wawona Hotel stands today. The **Tenaya Basin** and **Tenaya Canyon** were also entombed. On the east escarpment of the Sierra, the ice reached down into the **Mono Basin**. Vestigial glaciers remain atop **Mount Lyell**, **Mount Dana**, and **Mount Maclure** – the last remnants of the great icecap that sculpted the central Sierra.

Wherever glaciers flowed, rock eroded. Glaciers widened and deepened valleys, creating distinctive, U-shaped profiles,

Mono Lake

There is a sere majesty to Mono Lake, set in the high desert like a sapphire in a silver platter. It lies due east of Yosemite – a million years old, three times saltier than the ocean, and 80 times as alkaline. Its high mineral content is anathema to fish, but vast quantities of brine shrimp and alkali flies breed here. These attract flocks of migratory shorebirds and marine birds.

Mono Lake is astoundingly salty because it has no outlet. Water flows in but can exit only by evaporation, leaving behind salts and other mineral compounds. One of its most remarkable geological features is tufa towers, columns of calcium carbonate formed by submerged springs. The lake has grown steadily more mineralized since it last overflowed 60,000 years ago.

Tufa formations (above), slowly deposited by warm springs, are reflected in Mono Lake's briny surface.

The story of Mono Lake belongs to the Sierra Nevada. As the Sierra escarpment rose, the Mono Basin subsided. During the past four million years, the basin has sunk while filling with alluvial debris. At the same time, the Sierra has risen until its crest stands 7,000 feet above the lake.

As with Yosemite, glaciation was important, but glaciers didn't grind out the lakebed. Instead, meltwater flowing from Sierra glaciers filled, then overfilled, the lake. At maximum, Mono Lake was more than 800 feet deeper than it is now and almost three times larger.

Volcanism was also significant. As the Mono Basin subsided, the crust thinned under it, letting basaltic magma penetrate from the mantle. Over several million years, a huge reservoir of molten rock formed just south of the basin, now known as the Long Valley magma chamber. It erupted cataclysmically 730,000 years ago, expelling 150 cubic miles of ash and rock into the atmosphere. Up to 600 feet of ash covered portions of the Mono Basin. The chamber is now believed to be almost fully recharged with magma.

A magma chamber of similar size exists between the Mono Basin and Long Valley, directly beneath a series of dormant vents called the Mono Craters. These erupted between 100,000 and 40,000 years ago – just yesterday in geologic time. And smaller magma chambers lie under Mono Lake itself, forming two islands, Paoha and Negit.

Given the quantity of magma in the region, volcanism will surely affect Mono Lake's future. Some geologists think the next act will begin sooner rather than later: The Long Valley Caldera is often shaken by earthquakes, a sign that magma is creeping toward the surface.

saddle-shaped passes known as cols.

Perhaps the best place in Yosemite to observe the parts of a glaciated valley is the **Lyell Fork** of the Tuolumne River, a long, lovely wilderness valley south of **Tuolumne Meadows**. From vantages near the headwaters of the Lyell Fork on the slopes of Mount Lyell, the entire geologic history of the valley becomes as explicit as an illuminated text.

Yosemite Valley was also filled with ice and demonstrates glacial features such as hanging valleys. These are side-valleys unlike the sharp, V-shaped valleys carved by rivers alone. Cirques – bowl-shaped depressions – formed at the head of the glaciers. These were backed by once-smooth cliffs that eventually developed sharply crenellated ridges, or arretes, at the tops of the headwalls, along with peaks called horns, and formed by tributary streams that once connected with the primary river. They were literally left hanging when the main river valley was scoured deep by glaciers. **Bridalveil Fall**, across from El Capitan, pours out of a hanging valley and drops 620 feet. Farther east on the northern side of the valley,

Yosemite Falls goes this one step better: It plunges 1,430 feet, passes through a channel 815 feet high, then drops a final 320 feet, making a total descent of over 2,500 feet.

But perhaps the most widespread evidence of glaciation in Yosemite is what the rivers of ice left behind. As glaciers move, they carry boulders, rocks, sand, and soil, collectively known as till – an apt term, given that glaciers act as gargantuan plows. Sometimes the till is deposited as a layer of rocky material as the glacier moves along. Sometimes big boulders are dropped in unlikely places, such as atop larger boulders or on sheets of rock. These are known as glacial erratics. But the most distinctive evidence of a glacier's passage are moraines: long ridges composed of till that occupy glacial valleys. Moraines are abundant throughout the park and can be easily observed, even in timbered areas. Look for sharp ridges with jumbled boulders at the crests.

When the Tioga glaciation ended about 10,000 years ago, the glaciers retreated and Yosemite Valley filled with water, creating the now vanished Lake Yosemite.

This must have been strikingly beautiful, framed by the soaring cliffs of El Capitan and Half Dome. Lake Yosemite gradually dried up, but rivers and streams continued to flow over the terrain vacated by the glaciers, redistributing the till. This created a seedbed where grasses, sedges, and perennials took root, resulting in the meadows that are now one of the park's most pleasant features.

To the casual visitor, Yosemite may seem like a finished work, a set piece of nature's most spectacular scenery. But the park is as dynamic as ever. Granite slabs still exfoliate, and the Merced and Tuolumne Rivers continue to scour their canyons deeper. We're watching only the latest act in a grand geologic play that began eons ago and will stretch eons into the future.

Staircase Falls (right) trickles down about 1,300 feet over a series of granite ledges.

Erratic boulders (below) were transported to this rocky slope by glaciers about 10,000 years ago.

TRAVEL TIPS

DETAILS

When to Go

Avoid summer crowds by visiting in late spring or early fall. If you must come in summer, reserve lodgings six months to a year in advance. Summer highs in Yosemite Valley sometimes reach 95°F; expect thunderstorms. Winter is relatively mild in the valley, but snowfall is heavy at higher elevations. Weather can change suddenly at any time of year, so be prepared for a wide range of conditions.

How to Get There

The nearest airports are Fresno Yosemite International Airport, a three-hour drive away, and Merced Municipal Airport, two hours away. Grayline of Yosemite (888-727-5287) offers year-round bus service from Merced and from Fresno in summer. Yosemite is 185 miles from San Francisco and 215 miles from Reno, Nevada.

Getting Around

Car rentals are available at both airports. Free shuttle bus service is available year-round in Yosemite Valley.

Handicapped Access

The visitor centers, nature center, and some trails and campsites are wheelchair-accessible.

INFORMATION

Yosemite National Park

P.O. Box 577, Yosemite, CA 95389; tel: 209-372-0265.

California State Division of Tourism

801 K Street, Suite 1600, Sacramento, CA 95814; tel: 800-462-2543 or 916-322-2881.

CAMPING

Yosemite National Park

National Park Reservation System, P.O. Box 1600, Cumberland, MD 21502; tel: 800-436-7275.

There are 13 campgrounds in the park, and more than 1,400 campsites. Some sites may be reserved in advance; others are first-come, first-served. A permit is required for backcountry camping. To obtain one, send a request to Wilderness Permits, P.O. Box 545, Yosemite National Park, CA 95389, or call 209-372-0740. Include your name, address, phone number, number of people, method of travel, start and finish dates, and intended trail route.

LODGING

PRICE GUIDE – double occupancy

$ = up to $49 $$ = $50–$99
$$$ = $100–$149 $$$$ = $150+

Ahwahnee

Yosemite Concessions Services, 5410 East Home Avenue, Fresno, CA 93727; tel: 559-252-4848.

This extraordinary, six-story structure was called by Ansel Adams "one of the world's distinctive resort hotels." Opened in 1927 and refurbished in 1979, it has 99 rooms, four suites, and 24 cottages. Situated about a mile east of Yosemite Village, the hotel offers views of Half Dome, Glacier Point, Royal Arches, and Yosemite Falls. $$$$

Curry Village

Yosemite Concessions Services, 5410 East Home Avenue, Fresno, CA 93727; tel: 559-252-4848.

In the shadow of Glacier Point, Curry Village offers 427 tent-cabins, 183 cabins, and 20 hotel rooms. Wood and canvas cabins hold as many as five and are provided with linens and blankets; bathrooms are shared and centrally located. Free-standing wood cabins, with or without private baths, accommodate up to five people. Amenities include a cafeteria, pool, ice-skating rink, gift shops, and bicycle rentals. Open spring to fall and on weekends and holidays in winter. $–$$

Wawona Hotel

Yosemite Concessions Services, 5410 East Home Avenue, Fresno, CA 93727; tel: 559-252-4848.

Set in Wawona, about 30 miles from Yosemite Valley on Highway 41, this Victorian-era hotel is the oldest in the park, built in 1856 and refurbished in 1983. There are 104 rooms furnished in period style; about half have private bathrooms. A dining room, lounge, and golf shop are on the premises. $$–$$$

Yosemite Fish Camp Bed-and-Breakfast Inn

1164 Railroad Avenue, P.O. Box 25, Fish Camp, CA 93623; tel: 559-683-7426.

Three guest rooms, one with private bath, are available in a house about two miles from the south entrance to the park. $–$$

Yosemite Lodge

Yosemite Concessions Services, 5410 East Home Avenue, Fresno, CA 93727; tel: 559-252-4848.

Affording good views of both Yosemite Falls and the Merced River, this lodge's 495 units offer three types of accommodations: deluxe rooms with balconies or patios, standard hotel rooms with or without a private bath, and rustic cabins with or without a private bath. A restaurant, lounge, pool, gift shop, and bicycle rentals are available. $$–$$$

Yosemite View Lodge

11136 Highway 140, P.O. Box D, El Portal, CA 95318; tel: 209-379-2681.

A few miles west of the park, the lodge has 280 rooms with kitchenettes. About half the rooms have spa tubs, fireplaces, and private balconies overlooking the Merced River. A restaurant, bar, coin laundry, gift shop, and indoor and outdoor pool are on the grounds. $$–$$$$

TOURS

Yosemite Association

P.O. Box 230, El Portal, CA 95318; tel: 209-379-2321.

Naturalists conduct dozens of seminars each year on such topics as geology, astronomy, and birding. The association also leads two- to seven-day backpacking excursions in the park. Call or write for a catalog.

Yosemite Concessions Services

Tour Desk, Yosemite National Park, Yosemite, CA 95389; tel: 209-372-1240.

Guides lead daily sightseeing tours in open-air trams and motor coaches. Tours last one to eight hours and focus on the park's geology, flora, and fauna. Guided backpacking tours also take small groups into the backcountry.

Yosemite Valley Stables

Yosemite, CA 95389; tel: 209-372-8348.

Tours can accommodate both novice and experienced riders for two-hour, four-hour, and all-day riding excursions on mule or horseback through Yosemite's beautiful backcountry; four- to six-day saddle trips through Yosemite's High Sierra; and open-ended pack trips in the park.

Excursions

Devils Postpile National Monument

47050 Generals Highway, Three Rivers, CA 93271; tel: 559-565-3134.
Less than 100,000 years ago, a lava flow poured down the Middle Fork of the San Joaquin River on the eastern flank of the Sierras. It cooled slowly enough to form columnar joints in the basalt. When glaciers crunched over the flow, they tore away part of it, leaving the Postpile and its 60-foot high cliff. Hike around the formation to see the columns and the glacial polish on top. Another hike leads to 101-foot-high Rainbow Falls, where the river drops over a cliff of andesite and rhyodacite.

Lassen Volcanic National Park

P.O. Box 100, Mineral, CA 96063; tel: 530-595-4444.
The eruption of Lassen Peak in 1914-1915 was the most violent volcanic activity in the United States before that of Mount St. Helens in 1980. Highlights in the park include the Devastated Area, a huge expanse of forest that was leveled by the 1915 eruption, and hikes up 10,457-foot Mount Lassen and a nearby cinder cone. The steaming mudpots and fumaroles at Bumpass Hell and the Sulphur Works testify to the presence of a geothermal hot spot that continues to blaze beneath the surface.

Sequoia and Kings Canyon National Parks

47050 Generals Highway, Three Rivers, CA 93271-9651; tel: 559-565-3134.
These adjacent parks in the southern Sierra Nevada are renowned for their sequoia forests, spectacular waterfalls, and glacially carved peaks and canyons. Most of the region is accessible only by trail, including a segment of the John Muir Trail which stretches more than 200 miles from Yosemite Valley to the crest of Mount Whitney, the highest peak in California. Exploring by car can take you to Cedar Grove, a glaciated valley almost as impressive as Yosemite and much less crowded.

San Andreas Fault
California

Point Reyes National Seashore is a beautiful wilderness of forested ridges, brackish bays, and vast beaches in the San Francisco Bay Area. Unfortunately for local residents, this recreational haven is headed for Alaska. It may take 40 million years, but the Point Reyes Peninsula will snap off from the California coast as it continues to hitch a ride northwestward atop the Pacific plate, the biggest puzzle piece in Earth's shifting crust. ◆ Blame the **San Andreas Fault**. The longest and most famous seismic scar in North America, the San Andreas rubs and slides between the Pacific and North American plates for 750 miles from the Salton Sea near the Mexican border to north of Point Delgada in Northern California. It's a textbook example of what geologists call a transform fault, and it creeps anywhere from half an inch to two inches per year. ◆ Before dawn on April 18, 1906, the San Andreas ruptured over a length of 300 miles, unleashing a 40-second earthquake and three days of fire that turned

Tremors ripple through the Earth as the California coast inches northward along the continent's longest fault.

San Francisco's post-gold rush boom to bust. The city built on gold dust was left in ashes and rubble. Extensive surveys after the earthquake discovered the fault and spawned a theory for why earthquakes occur, which experts still largely accept. According to this elastic rebound theory, strain builds up in the crust over decades, until the weakest section of rock fails suddenly and the two sides of the fault slide in opposite directions. The greater the strain, the stronger the quake.

The San Andreas Fault, powered by the collision of the Pacific and North American plates, cuts through the state like a surgical scar.

Tracing the Fault

"Where's the earthquake?" asks an impatient three-year-old walking the easy half-mile-loop **Earthquake Trail**, just east of the **Bear Valley Visitor Center** at Point Reyes National Seashore. While there's not much chance you'll feel an earthquake on the trail, you will see signs of the dramatic 1906 event. The largest horizontal displacement – 16 feet or more – occurred here in **Olema Valley**, beneath which runs the fault on its way to **Bolinas Lagoon**.

Local legend says the ground gaped open on the Shafter Ranch here in 1906, swallowed Matilda the cow, and closed up with only her tail protruding. Other accounts suggest that the already dead cow had been dumped into the trench by a ranchhand who then fed the more sensational story to overeager reporters. The truth, like the six-foot-wide rift that tore across the surface, has been obscured by the passage of time.

Just beyond the weathered bridge over

Urban Geology

Earthquakes are integral to San Francisco's identity. A major quake rocked the burgeoning city at least once a decade between the gold rush of 1849 and that fateful morning of April 18, 1906, when a magnitude 8.3 event toppled buildings and unleashed a three-day inferno that left four square miles in ashes.

Few traces of that tragedy remain, but several spots reveal chapters of San Francisco's eclectic geologic story. Under the city lies the Franciscan mélange, a conglomerate of sandstone, serpentine, and other rocks flung together during collisions of the Farallon and North American plates when dinosaurs roamed the continent. Five distinct Franciscan rock types cross the city in bands, like pancakes stacked on a plate, from the bay to the Pacific Ocean: Alcatraz terrane, Hunters Point mélange, Marin Headlands terrane, City College mélange, and San Bruno Mountain terrane.

The Alcatraz terrane consists of thick sandstones that form the hills favored for Hollywood car chases and Alcatraz Island itself. Long before The Rock became a notorious prison, it was just another of San Francisco's hills until the melting of glaciers filled the bay with water. A ferry trip to Alcatraz is a great way to view the bay and the sandstone; also try Telegraph Hill at Sansome and Union Streets.

For serpentine, California's state rock, in the Hunters Point mélange, head to **Fort Point National Historic Site**, beneath the southern end of the **Golden Gate Bridge**. As waves crash and traffic rumbles overhead, ponder the journey made by the prominent gray-green outcrop beside the fort. It once sat deep under the seafloor on the Farallon plate but has been squeezed up like toothpaste and scraped onto the edge of the continent by plate tectonics.

Across the Golden Gate from Fort Point, the **Marin Headlands** display textbook folded beds of red chert, another deep-ocean immigrant. You can see more of this signature rock of the Marin Headlands terrane from atop **Twin Peaks** in central San Francisco. These twins are hardly identical: The north peak consists of red chert, best exposed on the summit, while the amorphous rock of the south peak is pillow basalt. Having erupted on the Pacific seafloor, the basalt slowly migrated to North America. Like a lot of the city's people, San Francisco's rocks have settled here from far and wide and are still on the move.

Folded beds of chert (top, left) testify to the intense tectonic pressures that shaped the rocks underlying the Marin Headlands (below).

City Hall (opposite) was one of the first structures to collapse during the 1906 earthquake; it had just been completed after three decades of construction.

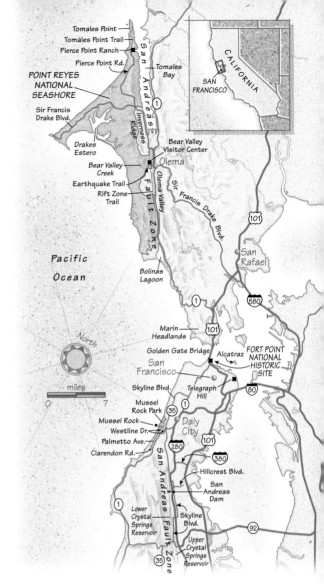

Bear Valley Creek, a line of blue posts across the hillside follows the faultline, visible as a narrow ledge of earth. A big red barn nearby straddles the fault; its predecessor was knocked 16 feet off its foundation in 1906. Preserved along the trail is a picket fence perpendicular to the fault, which split and shifted by an equal distance. On the uphill side you can stand astride the faultline, one foot firmly planted on granite transplanted from the southern Sierra Nevada and the other on Franciscan rock from the mid-Pacific.

These two kinds of bedrock beneath Point Reyes attest to the slow march of plate tectonics. This is a land in motion, mostly imperceptible yet inexorable. Boulders beside the trail let rock hounds examine the granite, sandstone, and serpentine from the two sides of the fault.

A short walk down the nearby Bear Valley Trail leads to the **Mount Wittenberg Trail,** which winds for two mostly uphill but well-shaded miles to the 1,400-foot summit of this granite massif. Made of rock that formed in Southern California, Mount Wittenberg owes its present location to the movement of the San Andreas. From the summit, the dramatic view to the south follows the rolling relief of the fault zone.

For a longer scenic hike in quake country, go three miles south of the town of **Olema** on Highway 1, through the fault zone, past rolling hills and horse ranches, and turn right at Five Brooks Stables. Arrange a horseback ride or simply hoof it yourself to the **Rift Zone Trail,** a moderate 4.4-mile walk in the shade of conifers, laurels, and oaks, back toward Bear Valley. If you do walk, wear proper boots – the trail can be muddy and rutted.

Part of the trail follows a narrow ridge formed by the gradual grinding of the fault. Sunken land beside the trail contains sag ponds: chocolate-brown bogs fringed with ferns that are created by fault motion. Older ponds have filled in to become meadows. Watch for subtle changes in soil. On either side of the second creek crossing the trail, sandy yellowish patches and gray sandstone cobbles signal Franciscan bedrock, less resistant to the cutting action of creeks.

Tectonic Benefits

Skirting the edge of **Tomales Bay,** a long, linear body born and bisected by the fault, Sir Francis Drake Boulevard sits in the shadow of lushly forested **Inverness Ridge,** a granite monolith that migrated more than 300 miles north from the Tehachapi Mountains. Across the bay, bare brown, rounded hills rise like fists, their slopes incised by creekbeds. This eastern side of the bay and the fault consist of Franciscan conglomerate.

Approaching the north tip of the Point Reyes Peninsula on Pierce Point Road, you pass working dairies and have spectacular views of the headlands and Tomales Bay. The 4.7-mile **Tomales Point Trail** begins from historic Pierce Point Ranch and follows the spine of the peninsula high above the Pacific's crashing waves toward the mouth of Tomales Bay, where the greatest ground displacement occurred in 1906.

Watch the ground for the occasional boulder or outcrop of granite. At a saddle in the trail, you can look east beyond Tomales Bay or west across sunlight-stippled waves. While peering down sheer sea cliffs to a deserted, beckoning beach, consider that without the San Andreas fault, California wouldn't have much of its spectacular scenery. Uplift along the fault helped sculpt the precipitous coastline and the more rounded contours of the inland Coast Ranges. It scooped out that bountiful breadbasket, the Great Central Valley. Seismologist Allan Lindh of the U.S. Geological Survey argues that the contribution from the San Andreas to California's topography and economy provides up to ten times the income that the fault removes by its periodic tectonic throes.

Seismic Neighborhoods

After passing beneath the ocean west of San Francisco, where the 1906 quake's epicenter was located, the fault emerges onshore at **Mussel Rock** in **Daly City**. A massive landslide just south of Mussel Rock marks the San Andreas fault zone, and seismic activity is crumbling the bluff close to homes built on top. To glimpse the action, take the Manor Drive/Palmetto Avenue exit from Highway 1, follow Palmetto north along the coast, and turn left on Westline Drive. Park where Westline makes a sharp uphill bend to the east.

Wavelike bulges ripple across the road, and wide cracks have thwarted repeated attempts at patching. One crack can be traced from one side of the street to the other, across a yard, and into a house foundation. The north side of Westline here has slumped downslope two feet – its sidewalk broken and buckled. Just beyond a fence lies the landslide scar.

From here on the south side of the fault zone, you look across to the cliff face situated on the north side. Talus cones of debris from the friable Franciscan rocks trickle downslope. Pink-and-white houses appear to loom precariously close to the edge, which some years recedes as much as three feet. The fault itself remains obscured by landfill and overgrown debris, but frequent low-flying jets seem to follow its trace as they head out over the Pacific and turn north toward Point Reyes and Mount Tamalpais.

An Unshakable Dam

From Mussel Rock, the San Andreas continues southward for another 500 miles. To see the region that lent its name to the fault, take Highway 280 south to

Anatomy of an Earthquake

The Earth threw the first pitch in the third game of the 1989 World Series. At 5:04 P.M. on October 17, as the Oakland A's and San Francisco Giants prepared to play ball at San Francisco's Candlestick Park, the game was called on account of an earthquake – surely a first for this or any sport. Fortunately, the stadium stood on solid rock, but the magnitude 7.1 Loma Prieta earthquake still wreaked more than $6 billion in damage around Northern California. Remarkably for a rush-hour temblor, only 67 people died.

Centered about 60 miles south of San Francisco in the **Santa Cruz Mountains**, the 20-second quake began 12 miles deep in the Earth's crust. It displaced the fault with a four-foot vertical jolt, in addition to the sideways slip typical along the San Andreas. A 25-mile-long segment broke, but the rupture failed to reach the surface as it spread north and south at more than a mile per second.

Tangible evidence that this had been a major earthquake covered a 3,000-square-mile area. The Loma Prieta event crippled downtown Santa Cruz's shopping arcade, crumpled a mile-long, two-story highway in Oakland, and dropped a span of the San Francisco-Oakland Bay Bridge like a massive trapdoor. In San Francisco's trendy South of Market and pricey Marina districts, buildings sank, slipped, and collapsed as the strong ground motions turned the landfill beneath them to jelly.

Considering their distance from the quake, it was surprising that the cities competing in the cross-bay World Series suffered major damage. As surprising was the capricious distribution of the damage. In Oakland, the collapse of the I-880 Cypress structure caused most of the deaths, while buildings beside the highway emerged unscathed. Shock waves from the unusually deep origin may have bounced off deeper, denser rock in the crust and been reflected to the surface on either side of San Francisco Bay. Though this wasn't the Big One that Bay Area residents have long feared, it was big enough.

The collapse of an Oakland freeway caused most of the deaths during the 1989 Loma Prieta quake.

A fence along Point Reyes' Earthquake Trail (opposite) was wrenched about 16 feet apart during the 1906 quake.

Valley. A plaque on a serpentine boulder atop the dam marks where the fault trace passes and continues along the lake's eastern shore.

South from the dam, a tranquil, tree-filled valley extends to **Crystal Springs Reservoir**. Under creaking bay laurels, the easy trail resumes

the Larkspur Drive/Millbrae Avenue exit, turn right on Skyline Boulevard and again on Hillcrest, which comes to a dead end at the **Sawyer Camp Trail**. In just under a mile, this winding, paved path leads to the San Andreas Dam. When completed in 1869, the dam filled part of a linear, 15-mile-long valley formed by the fault. The dam withstood the shaking in 1906, although its eastern abutment shifted nine feet to the northwest.

Measurements of ground motion here before and after the 1906 earthquake showed the existence of the fault, which geologist Andrew Lawson named for the San Andreas

from the west end of **San Andreas Dam** and reaches a branch of Crystal Springs Reservoir in under two miles. Other trails nearby provide views of the fault zone, and the San Andreas leaves its mark elsewhere throughout central and Southern California.

Indeed, the fault is really a network of several distinct branches. One of them, the Hayward fault at the base of the Berkeley Hills, is more likely than the San Andreas to trigger the next major earthquake in the San Francisco Bay Area. If it approaches the 1906 temblor in magnitude, the next Big One won't soon be forgotten.

TRAVEL TIPS

DETAILS

When to Go

Summer in the Bay Area is warm inland, with temperatures in the 80s, but cool and often foggy along the coast, with temperatures in the 50s and 60s. June through September is the dry season, with little or no rain; November through March tends to be cool and rainy. Average daytime temperatures in winter are in the 50s.

How to Get There

Commercial airlines serve San Francisco and Oakland, about two hours east and south of Point Reyes.

Getting Around

A car is a necessity for exploring Point Reyes and the surrounding area. Rentals are available at the airports.

Handicapped Access

The visitor centers, the Earthquake Trail at Bear Valley, and a paved trail at Limantour Beach are accessible. A wheelchair with large balloon tires, suitable for use on beaches and unpaved trails, is available by reservation from the Point Reyes National Seashore Association; call 415-663-1200.

INFORMATION

Marin County Convention and Visitors Bureau

1013 Larkspur Landing Circle, Larkspur, CA 94939; tel: 415-499-5000.

Point Reyes National Seashore

Bear Valley Visitor Center, Point Reyes Station, CA 94956; tel: 415-663-1092.

The Ken Patrick Visitor Center is at Drakes Beach; Lighthouse Visitor Center is at the peninsula's western tip.

West Marin Chamber of Commerce

P.O. Box 1045, Point Reyes Station, CA 94956; tel: 415-663-9232.

CAMPING

Permits are required for the national seashore's four hike-in campgrounds; check with the Bear Valley Visitor Center (415-663-1092). Visitors may reserve sites up to three months in advance by calling 415-663-8054. Samuel P. Taylor State Park, six miles southeast of seashore headquarters, offers the closest drive-in camping; call 415-488-9897 for information and 800-444-7275 for reservations.

LODGING

PRICE GUIDE – double occupancy	
$ = up to $49	$$ = $50–$99
$$$ = $100–$149	$$$$ = $150+

Bay View Cottage

P.O. Box 638, Point Reyes Station, CA 94956; tel: 415-663-8800.

Set on two hilltop acres near the national seashore, this one-bedroom country cottage has a fireplace, sunny deck with hot tub, galley kitchen, and views of Tomales Bay. $$$

Berry Patch Cottage

P.O. Box 712, Point Reyes Station, CA 94956; tel: 415-663-1942 or 888-663-1942.

This one-bedroom cottage is set in a secluded glen just outside the park and has a deck, garden, and fully equipped kitchen. $

Dancing Coyote Beach

P.O. Box 98, Inverness, CA 94937; tel: 415-669-7200.

Guests stay in four cottages overlooking Tomales Bay. Each has a fireplace and a full kitchen stocked with supplies for breakfast. $$$–$$$$

Golden Hinde Inn

P.O. Box 295, Inverness, CA 94937; tel: 415-669-1389 or 800-339-9398.

The inn has 35 rooms, some with private baths, fireplaces, and/or views of Tomales Bay; eight units have kitchenettes. A restaurant and marina are on the premises. $$–$$$

Holly Tree Inn

Box 642, Point Reyes Station, CA 94956; tel: 415-663-1554.

Guests stay in four rooms in the main house or in a two-room cottage tucked at the edge of the large garden. A country breakfast is served, and a hot tub is on the premises. The inn has two other cottages: Sea Star, at Tomales Bay, and Vision, located inland. $$$–$$$$

Hotel Inverness

25 Park Avenue, Inverness, CA 94937; tel: 415-669-7393.

The hotel is in a 1906 shingle-style building a block from Tomales Bay in the historic section of Inverness. The five guest rooms all have private baths; two rooms have a private deck and two have sitting areas. Breakfast is served in the rooms or on an outdoor deck. $$$–$$$$

Manka's Inverness Lodge

P.O. Box 1110, Inverness, CA 94937; tel: 415-669-1034.

Built originally as a hunting lodge, this small compound includes two cabins and 11 guest rooms, as well as a gourmet restaurant. Also for rent is an 1850s cottage on Tomales Bay. $$$–$$$$

Neon Rose

P.O. Box 632, Point Reyes Station, CA 94956; tel: 415-663-9143 or 800-358-8346.

Adjacent to the national seashore, the Neon Rose is a fully equipped guest cottage, with skylights and views of Tomales Bay. It has a fully equipped kitchen, living room with fireplace, full bath with Jacuzzi, and a private garden. $$$–$$$$

Olema Inn

10000 Sir Francis Drake Boulevard, Olema, CA 94950; tel: 415-663-9559.

This two-story inn, built in 1876, has six guest rooms with private baths. A restaurant is on the premises, and breakfast is included in the rate. $$$

Point Reyes Seashore Lodge

10021 Coastal Highway 1, Box 39, Olema, CA 94950; tel: 415-663-9000.

This bed-and-breakfast lodge has 21 rooms with private bathrooms, including three suites. Many of the rooms have fireplaces, and most have whirlpool tubs. Two cottages, each sleeping up to four people, have their own kitchens, private decks, and Jacuzzis. $$$–$$$$

TOURS

Point Reyes National Seashore

Bear Valley Visitor Center, Point Reyes Station, CA 94956; tel: 415-663-1092.

Park rangers and naturalists offer guided tours and field seminars throughout the year; some focus on the geology of earthquakes in the Point Reyes area.

Excursions

Big Sur

Big Sur Chamber of Commerce, P.O. Box 87, Big Sur, CA 93920; tel: 831-667-2100.

Coastal California is a geological wrecking yard. As tectonic forces drove pieces of the Pacific Oceanic plate against North America, ocean sediments were crushed and smeared onto the edge of the continent. In some places, deep basement rocks have been brought to the surface by erosion and faulting; elsewhere, fragile Franciscan "mélange" – soft, friable rocks composed of marine sediments – are exposed. Scenic Point Lobos near Carmel is granite and diorite from deep down, while the coast ranges, including Point Sur itself, are mostly Franciscan rubble. As you travel, look inland for flat-lying ledges that mark former shorelines, now lifted high above the sea.

Pinnacles National Monument

5000 Highway 146, Paicines, CA 95043; tel: 831-389-4485.

Rugged, spirelike formations rise 500 to 1,200 feet above the smooth hills of the countryside southeast of Salinas. The rocks are the remains of an ancient volcano whose other half lies almost 200 miles southeast, thanks to the San Andreas fault. Explore the hiking trails in the East and West Districts, which are unconnected by a direct road through the park.

Lava Beds National Monument

P.O. Box 867, Tulelake, CA 96134; tel: 530-667-2282.

Northeast of Mount Shasta, almost at the Oregon border, volcanic eruptions on the Medicine Lake shield volcano have built a rugged landscape of lava flows, lava tubes, cinder cones, and pit craters. The lava tubes, broken open where their ceilings have collapsed, are about 30,000 years old and form a multitude of basaltic caves which can be explored on foot. Some of the lava-tube caves contain year-round deposits of ice. Some cinder cones are only about 1,000 years old.

Mount St. Helens

Washington

CHAPTER **19**

ittle prepares you for your first sight of Mount St. Helens. Approaching from the west, **Spirit Lake Memorial Highway** winds along the **North Fork Toutle River** beneath hills blanketed by second-growth fir. Alders, maples, and cottonwoods are tossed by the breeze along the river's edge. But as you crest Hoffstadt Bluff, 21 miles past the **Mount St. Helens Visitors Center**, the valley opens into a primal landscape – stark, strange, and beautiful. ◆ Fifteen miles to the southeast, the shattered hulk of Mount St. Helens dominates a bare and broken topography. Beneath you, the river rushes through the stones and gravels of its rubble-filled bed, cutting sharply into banks of avalanche debris. You are still three miles downstream from the area leveled by the mountain's violent blast. But evidence of its awesome force is visible and close at hand. ◆ On the morning of May 18, 1980, Mount St. Helens awakened after a 123-year slumber with an explosive yawn. The blast was 27,000 times more powerful than a Hiroshima-scale atomic bomb.

A ravaged landscape bears witness to the "Fire Mountain's" explosive temper.

It leveled miles of forest and killed 57 people. Volcanic mudflows swept down the Toutle and Cowlitz Rivers, obliterating roads, buildings, and bridges, and blocking shipping on the Columbia River 70 miles from the mountain. The resulting ash cloud circled the globe. ◆ Prior to 1980, Mount St. Helens was considered the loveliest of Cascade volcanoes. Unmarred by Ice Age erosion, its slopes rose gracefully more than a mile above the shores of Spirit Lake and surrounding forests. To Northwesterners, it was the Fujiyama of the Cascades. But to the Klickitat Indians, whose oral traditions may recall a similar eruption 2,500 years ago, it was Tah-one-lat-clah, "Fire Mountain."

The eruption of Mount St. Helens in 1980 turned a beautifully forested part of the Cascade Mountains into a lunar landscape of ash and lava.

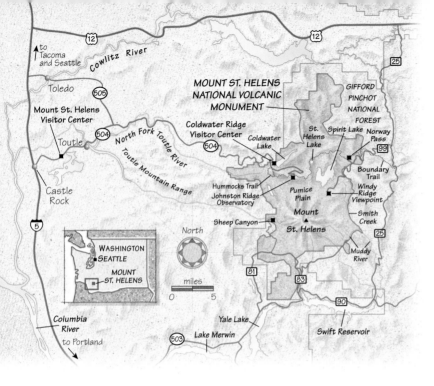

As ocean floors scrape beneath them, continental edges weaken and fracture. The seafloor basalts melt in the hot, semisolid mantle. Charged with water and gases from the melted seafloor and sediments, magma pushes up through faults and weaknesses in the continental crust. Where it breaches the surface, a volcano erupts.

The Cascade volcanoes bear witness to this process, but none so dramatically as Mount St. Helens. As magma pushed up into the volcano in the weeks before the eruption, its northern flank bulged visibly outward, steepening and fracturing the overlying rock. Geologists determined that the bulge was expanding at the astounding rate of five feet per day. A week before the eruption, it had swelled outward nearly 500 feet, like a geological aneurysm ready to burst.

Fire in the Belly

Like its loftier neighbors – Mount Rainier to the north, Mount Adams to the east, and Mount Hood to the south – St. Helens is a composite volcano formed of layers of lava and tephra (airborne rock and ash). It and all the Cascade volcanoes are part of the "Ring of Fire," a chain of more than 500 active volcanoes that circles the Pacific Ocean like a string of fiery pearls.

These volcanoes trace subduction zones where basaltic plates of the Pacific Ocean floor plunge under the lighter granitic continents at the rate of an inch or two per year.

On the morning of May 18, David Johnston, a volcanologist with the U.S. Geological Survey, was at an observation post on a ridge five miles north of the volcano. Early that morning, he radioed his latest measurements to the Cascade Volcano Observatory in Vancouver, Washington. At 8:32 A.M., his second transmission was cut short after five hurried words: "Vancouver, Vancouver. This is it!"

A magnitude 5.1 earthquake collapsed the bulging north face of the mountain. A massive debris avalanche swept down into Spirit Lake and the North Fork Toutle River, burying the landscape beneath hundreds of feet of debris. As the mountainside fell away, it exposed the highly charged magma, igniting an eruptive blast. Superheated gas and rock-laden winds of 400°F to 600°F shot over forested ridges at speeds initially exceeding 600 miles per hour. Within minutes, 230

square miles of forest north of the volcano were leveled. Ash and pumice rained down over the devastated landscape. No trace of Johnston was ever found.

Exploring the Aftermath

With the establishment of the **Mount St. Helens National Volcanic Monument**, the U.S. Forest Service developed several interpretive centers close by the mountain. There is much to see on the roads and trails that wind around the volcano, but an excellent place to start is the **Coldwater Ridge Visitors Center**, 38 miles up the Spirit Lake Memorial Highway. Situated within the devastated area, it interprets the unfolding story of ecological recovery at St. Helens. Just shy of the center, a turnoff leads down into the rubble-filled Coldwater Creek drainage and up to the ridge where David Johnston kept his final watch. Here, at **Johnston Ridge**

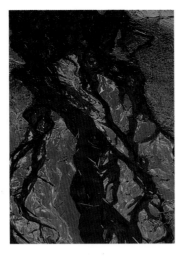

Trees around Spirit Lake (opposite) were scattered like toothpicks by the eruption.

Volcanic ash (left) choked the rivers, which have established new drainage patterns on top of the debris.

A series of small-scale eruptions after the main blast have built a lava dome in the center of the crater (below).

Observatory, a few miles from the dome, you can follow the geologic story of the mountain´through a film and interpretive exhibits. Outside, a short walk leads to a ridgetop overlook.

The view of the volcano and its transformed landscape is spellbinding. Clouds cling to ash-gray slopes, and snow fingers trail from remnant glaciers. The yawning chasm of the crater opens toward you in a

Hummocks Trail. This two-and-a-half-mile loop winds through the heart of the debris avalanche, a chaotic landscape formed by the massive landslide that triggered the eruption.

Klimasauskas points out that over 90 percent of what's missing from the volcano lies here in the valley. "One way to envision it," he explains, "is to imagine taking a cubic-mile ice cream scoop out of the old St. Helens, dropping it into a food processor, hitting 'pulse' a couple of times, and dumping it out."

The scooped-out hollow is the crater, and the pulverized mix is a chunky blend of the volcano's former summit and core: the hummocks. These 20- to 200-foot bumps are pieces of the pre-eruptive mountain. "Some are still intact, though badly fractured," Klimasauskas points out. "What's amazing is that they were carried six or seven miles and 5,000 feet in elevation from the summit to here in a matter of minutes."

Colors of Destruction

The first thing that strikes you in this jumbled landscape are the pastel hues of the deposits. Some are pale pink grading to buff and yellow, others are darker, rusts and oranges and browns swirled with black. And there are a thousand shades of gray.

The reds and blacks are the weathered surfaces of older lava flows, some having erupted more than 2,500 years ago. The lighter pastels, the yellows, light pinks, and oranges, are deposits that have been altered by hot water and acidic gases circulating through the summit dome. The weakening of these layers contributed directly to the magnitude of the 1980 collapse. In fact,

vast amphitheater. Its avalanche-streaked walls drop a thousand feet to the crater floor where a chunky dome of new lava is rising. A rubble-filled breach spills north from the crater onto a gently sloping plain. Deposited in a rain of hot, frothy rock in the first hours of the eruption, the **Pumice Plain** is now gray-green with new plant growth. Through it all the youthful rush of the upper Toutle River is carving a new watershed 700 feet above its old valley.

Ed Klimasauskas, a geologist with the U.S. Geological Survey, spends part of his time at Johnston Ridge doing research and helping visitors understand the origins, dynamics, and hazards of an active volcano. He has led groups throughout the monument, but his favorite place to experience the volcano is a few miles back down the road to the geologists say, one of the big lessons here is the hazard these same kinds of altered rocks pose in other Cascade volcanoes.

Here and there are cantaloupe-size rocks, light gray and slightly porous. These are newly formed chunks of dacite, lava from the 1980 summit dome. They have smooth, platelike faces where they cooled and cracked and rough "bread crust" outer surfaces, much like the newly forming summit dome. The new dacite is distinguishable from the older dacite and andesite lavas because the 1980 eruption didn't give time for crystals to form. A piece of older dacite from deeper in the mountain will have cooled more slowly; it will be speckled with flecks of black and white crystals.

The trail curls along the bases of jumbled hummocks, skirting small, willow-crowded ponds and wetlands that have formed between them. The steep, unstable hillsides

Fallen trees litter Windy Ridge (opposite, left).

Rocky hummocks (opposite, right) line the banks of the North Fork Toutle River, flung here by the violence of the eruption.

The explosion that formed Crater Lake (below) was many times greater than that which devastated Mount St. Helens.

are visibly eroding, but the flats between are green with moss, spreading plants, and small trees. The air is alive with bird songs.

Not far from the trail head, Ed Klimasauskas stops on a rise where the **Boundary Trail** branches off to ascend Johnston Ridge. To the east, bumpy hummocks stretch two miles to the slopes of the Pumice Plain. Three miles beyond that lies the stark hollow of the crater.

"No matter how many times I come here, the power and magnitude of this event humble me," Klimasauskas says. "I know it's happened several times before, but to have the chance to see it like this, fresh and unsettled – that's a rare privilege."

A Peaceful Lake Born of Cataclysm

Two hundred fifty miles south of Mount St. Helens, the ice-blue waters of a still, mountain lake mask one of the most violent volcanic eruptions in the continent's history. High in Oregon's southern Cascades, **Crater Lake** fills a volcanic caldera five to six miles across and nearly 4,000 feet deep. Steep, folded slopes drop from the crater rim to the lake's mirrorlike surface, its indigo blue deeper than sky. Beyond the rim, mature forests of pine and fir mantle the lower slopes of what was once a large mountain that dominated this landscape.

Geologists call it **Mount Mazama**. Mazama was a broad composite volcano; its clustered summit cones reached an estimated height of 10,000 feet. Like Mount St. Helens, it was fueled by the relentless jostling of the Earth's tectonic plates. And like St. Helens, it erupted violently.

Mazama exploded cataclysmically 7,000 years ago, sending a thick blanket of ash sprawling across 500,000 square miles, blackening the land as far east as Saskatchewan. Searing debris flows swept 25 to 40 miles down the surrounding valleys. In fact, so much magma erupted (more than 10 times that of Mount St. Helens) that the entire summit dome, 2,000 to 3,000 feet of the upper mountain, was blown off or collapsed, and the caldera was formed.

Over the next thousand years, smaller eruptions built the cinder cone of **Wizard Island**, a volcano-within-a-volcano that rises 700 feet above the surface of the lake. Rain and winter snowmelt slowly filled the caldera, and forests and wildlife reclaimed the mountain slopes. A road encircles the crater rim and offers turnouts to stunning overlooks. The short hike from **Crater Lake Lodge** to the 8,000-foot summit of **Garfield Peak** provides an intimate glimpse of lava flows, pumice, and volcanic debris – plus a sweeping view of the fire mountain's former domain.

TRAVEL TIPS

DETAILS

When to Go

July and August are relatively dry, with highs in the mid-70s. Fall and spring tend to be cold and rainy. Winter highs are in the 30s; heavy snowfall usually closes high-elevation roads from late October to April. The Mount St. Helens and Coldwater Ridge Visitor Centers are open year-round.

How to Get There

Commercial airlines serve Portland, Oregon, about 75 minutes away from the Mount St. Helens Visitor Center, and Seattle-Tacoma, Washington, about two hours away.

Getting Around

A car is the most convenient means of travel. Rentals are available at the airports.

Handicapped Access

The visitor centers and viewpoints are wheelchair-accessible. A few trails, including Big Creek Falls, Lava Canyon, Trail of Two Forests, Whistle Punk, and Woods Creek, are also suitable for wheelchairs; contact the park for details.

INFORMATION

Castle Rock Chamber of Commerce

147 Front Avenue Northwest, P.O. Box 721, Castle Rock WA 98611; tel: 360-274-6603.

Johnston Ridge Observatory

3029 Spirit Lake Highway, Castle Rock, WA 98611; tel: 360-274-2140.

Located on Highway 504 close to the mountain, the observatory is open from May to November.

Mount St. Helens National Volcanic Monument

42218 Northeast Yale Bridge Road, Amboy, WA 98601; tel: 360-247-3900.

Northwest Interpretive Association

3029 Spirit Lake Highway, Castle Rock, WA 98611; tel: 360-274-2127.

CAMPING

Gifford Pinchot National Forest

10600 Northeast 51st Circle, Vancouver, WA 98682; tel: 360-891-5000 or 877-444-6777 (reservations).

The national forest has more than 40 campgrounds. Iron Creek, with 98 sites, is the most popular.

Lewis and Clark State Park

4583 Jackson Highway, Winlock, WA 98596; tel: 800-452-5687.

The park, on Highway 12 northwest of the monument, has 25 campsites.

Seaquest State Park

3030 Spirit Lake Highway, Castle Rock, WA 98611; tel: 800-452-5687.

More than 90 sites are available at this park on Highway 504 near Silver Lake.

LODGING

PRICE GUIDE – double occupancy

$ = up to $49 $$ = $50–$99
$$$ = $100–$149 $$$$ = $150+

Blue Heron Inn Bed-and-Breakfast

2846 Spirit Lake Highway, Castle Rock, WA 98611; tel: 800-959-4049.

This seven-room inn is just across from the Mount St. Helens Visitor Center and offers views of Silver Lake and the volcano. Lodging includes a full-course dinner. $$$–$$$$

Crest Trail Lodge

12729 Highway 12, P.O. Box 490, Packwood, WA 98361; tel: 800-477-5339.

The new inn, situated between Mount St. Helens and Mount Rainier, offers 27 rooms with private baths. Many rooms are equipped with a refrigerator, cable television, and air-conditioning; two have wheelchair-accessible bathrooms. A continental breakfast is served. $$

Hotel Packwood

104 Main Street, P.O. Box 130, Packwood, WA 98361; tel: 360-494-5431.

This inexpensive hotel, built in 1912 about halfway between Mount St. Helens and Mount Rainier, offers nine simply furnished rooms, two with private bathrooms. $

Mount St. Helens Motel

1340 Mount St. Helens Way, Castle Rock, WA 98611; tel: 360-274-7721.

This 32-room motel offers simple accommodations, laundry facilities, and free morning coffee. $$

Shepherd's Inn

168 Autumn Heights Drive, Salkum, WA 98582; tel: 360-985-2434 or 800-985-2434.

Set on 40 wooded acres within driving distance of both Mount St. Helens and Mount Rainier, this bed-and-breakfast has five rooms, three with private bathrooms and a double Jacuzzi. Lodging includes a full breakfast. $$

Timberland Inn

1271 Mount St. Helens Way, Castle Rock, WA 98611; tel: 360-274-6002.

This modern motel offers 40 rooms and family suites. All units are equipped with refrigerators; some have Jacuzzis and microwaves. A coin laundry is on the premises. $–$$$

TOURS

C&C Aviation
14430 SE Center Street, Portland, OR 97236-2547; tel: 503-760-6969.

The company offers "flightseeing" tours of the volcano and surrounding area. The flight lasts 75 minutes and is limited to three passengers.

Mount St. Helens Adventure Tours
P.O. Box 149, Toutle, WA 98649; tel: 360-274-6542.

Four-hour tours explore the blast area in off-road vehicles and on foot.

Mount St. Helens National Volcanic Monument
42218 Northeast Yale Bridge Road, Amboy, WA 98601; tel: 360-247-3900.

Rangers lead tours to Ape Cave, a 1,900-year-old lava tube on the volcano's south side, and to Windy Ridge Viewpoint on the northeast side, with views of the crater and blast area. Tours occasionally visit other areas devastated by mudflows.

NW Discoveries
P.O. Box 23171, Tigard, OR 97281-3171; tel: 503-624-4829.

Experienced guides lead one-day hikes to the rim of the volcano. Transportation from the Portland area is available.

Excursions

North Cascades National Park
2105 State Route 20, Sedro Woolley, WA 98284; tel: 360-856-5700.

Route 20 (blocked by snow in winter) meanders across this rugged wilderness park of jagged peaks, deep valleys, waterfalls, and more than 300 glaciers. Geologically the region is a complex assemblage of "terranes," varied pieces of landscape pushed together by faulting and plate tectonics. Glaciers finished the job, sharpening mountain peaks, gouging out bowl-shaped depressions, and widening valleys. They continue to grind the rock into "glacial flour" which washes into the rivers and gives the water a turquoise hue.

Mount Rainier National Park
Tahoma Woods, Star Route, Ashford, WA 98304; tel: 360-569-2211.

Just 50 miles northeast of Mount St. Helens stands volcanic Mount Rainier, 14,410 feet high. Rainier's first eruption was about 500,000 years ago, its last in the 1840s. Though still considered active, its lava flows are largely buried under ice and snow. Several glaciers descend the mountain. Nisqually (on the south side) is the easiest to reach; its terminus is about a mile from the Henry M. Jackson Visitor Center on Highway 706.

Olympic National Park
600 East Park Avenue, Port Angeles, WA 98362; tel: 360-452-4501.

Scratch away the mantle of snow and ice atop Mount Olympus, and you'll find fossils of ancient sea creatures entombed in alpine rock. About 30 million years ago, as the Pacific plate collided with North America, the upper layers of the ocean floor were planed off and thrust upward, creating the jigsaw of basalt and sedimentary rocks that are now the Olympic Mountains. Glaciers continue to refine the landscape, fed by moist Pacific breezes that dump about 200 inches of precipitation on the crest each year. Highway 101 skirts the park boundary, and spur roads lead into the interior, but most of the park can be reached only on foot.

Channeled Scablands

Washington

CHAPTER 20

When the *Pathfinder* spacecraft landed on Mars in 1997, the world marveled at the beamed-back images of the alien planet. From orbital spacecraft, scientists had long seen that channels and grooves were etched into the Martian surface. Sure enough, *Pathfinder* touched down in a landscape strewn with boulders and other debris testifying to great floods in some distant epoch.   ◆   *Pathfinder*'s images were stunning, but they weren't a total surprise to the scientists who had prepared the mission. In an effort to learn what that part of Mars might look like close up, they had already studied the Channeled Scablands of eastern Washington, a region unlike any other on the planet. They had come because geologists today prize the scablands as the scene of dozens and maybe even a hundred Ice Age floods, the last perhaps only 15,000 years ago. From the evidence at **Grand Coulee** and **Dry Falls** to the curious craters, caves, and volcanic

Ice Age floods, bigger than all the world's rivers combined, scrawled a gigantic signature across the Northwest.

dikes dotting the wheat fields near the small town of **Odessa**, the region is a veritable textbook of catastrophic geology.   ◆   Several thousand feet above **Lincoln County**, Washington, it's a clear late-winter morning, and local farmer Tom Weishaar is showing visitors around in his four-seater Skywagon airplane. Most times, Weishaar goes airborne to survey the 1,800 acres of irrigated farmland he works north of Odessa, his fields arrayed amid another 4,000 acres of untillable scablands. But today he is taking his guests on a grand tour of this remote country.   ◆   First they fly up a well-defined channel where, one after another, small lakes remain from the Ice Age torrents: **Pacific Lake**, **Deer Lake**, **Coffeepot Lake**, and **Twin Lakes**.

Palouse Falls plunges 180 feet into a pool that was gouged out by the much larger floods that poured through the region toward the end of the Ice Age.

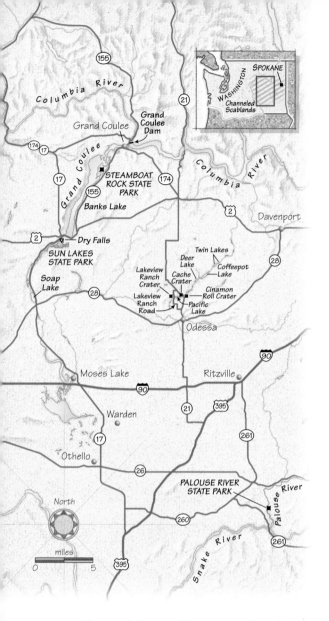

Aftermath of a Catastrophe

The story began thousands of years ago with a now-vanished body of water called Glacial Lake Missoula. This is the name geologists give to a meltwater lake that lay in northwestern Montana on the edge of the great continental ice sheet. Pent-up by a lobe of ice, the lake steadily deepened. One day the ice dam burst, loosing about as much water as Lakes Erie and Ontario combined. Geological evidence suggests that the cycle of fill-and-flood occurred repeatedly as the ice margin grew and shrank.

The floods swept through northern Idaho and onto the Columbia Plateau. Flowing southwest, they squeezed through Wallula Gap, just six miles north of today's Washington-Oregon border. From there, the floods turned west to scour the walls of the Columbia River Gorge, then south to inundate the present Willamette River Valley south of Portland. Everywhere you look in the Northwest, the evidence remains – from glacial-deposited boulders hulking in abandoned river channels to mile-long gravel bars rising from nowhere to graceful waterfalls flowing from ragged cliffs.

Although the scablands are best seen from the air, a driving tour gives visitors ample opportunity to imagine what such powerful floods may have been like. From Spokane – site of the nearest major airport – it's an 85-mile drive west on U.S. Highway 2 and state Highway 174 to **Grand Coulee Dam**. Built in the 1930s, the dam measures more than 5,000 feet long, 550 feet high, and 500 feet wide. Yet impressive as it is, the dam remains dwarfed by its surroundings. The Grand

Here and there, odd craters and earthen walls appear, along with circular patches of soil and scrubby vegetation atop basalt otherwise scraped bare – the very features that gave the scablands their name. The plane wings west toward Dry Falls, once the largest waterfall in the world, now a silent chasm. From this perspective, it's impossible to miss the floods' impact writ large on the land. And yet the Grand Coulee and scablands represent just part of the floods' legacy.

Coulee region's immense scale best reveals itself as travelers motor south from the dam.

Two-lane state Highway 155 hugs the **Columbia River**, here known as **Banks Lake** for the scenic stretch between Grand Coulee and Coulee City. At **Steamboat Rock State Park**, a steep, one-mile trail climbs from the campground to the 2,285-foot top of the rock. The slopes above the campground are littered with granite boulders that could only have been carried here by floodwater – "erratics," geologists call them, since they don't match the local basalt.

Just southwest of Coulee City on state Highway 17, the **Dry Falls Interpretive Center** in **Sun Lakes State Park** boasts both the single best view and most thorough explanation of the region's geological history. A painting shows what the area may have looked like during one of the great floods. The impact was such that Dry Falls, originally 20 miles to the south, retreated relentlessly upstream from the erosion. The walls of water probably raged 300 feet above the present canyon rim. Their flow was up to 10 times the combined flow of all the world's rivers. And when the floods finished, Dry Falls remained, an incomprehensible 3 miles wide and 400 feet high.

A small plaque outside the interpretive center pays homage to J Harlen Bretz, the geologist who first set forth the fantastic idea that this land had been carved by catastrophic floods. "Ideas without precedent are generally looked upon with disfavor and men are

Dry Falls (above), near Coulee City, is a scarp three miles wide and more than twice as high as Niagara Falls.

The Grand Coulee Dam (opposite) holds back the Columbia River.

A waterfall (right) in the Grand Coulee area trickles through landforms shaped by a series of Ice Age floods.

An Earthshaking Idea

It took scientists much of the 20th century to agree on just what formed the Channeled Scablands. Geologists once adhered to the school of uniformitarianism: that even the biggest and most unusual geological formations were the product of constant, uniform processes. But a renegade geologist named J Harlen Bretz (1882-1981) doubted that this theory could explain the tortured landscape of eastern Washington.

J Harlen Bretz taught at a Seattle high school and later was a professor of geology at the University of Chicago. He spent much of his free time traveling to the Grand Coulee and adjacent lands and poring over topographic maps of the region, and he was certain that the raw scablands and immense canyons must have been caused by great floods. By the mid-1920s, his theories were causing a sensation in the scientific world, but few of Bretz's peers believed him. They challenged him to explain where the huge floods had come from – and Bretz couldn't say.

Soon, however, the truth came trickling out. Scientists already knew about Glacial Lake Missoula, which covered hundreds of square miles in northwestern Montana during the Ice Age, held back by a massive ice dam on the Clark Fork River. Bretz suggested that this dam's failure could have unleashed a giant flood, and fellow geologist Joseph Pardee bolstered Bretz's conclusion with research on ripple marks – more evidence that floodwaters had repeatedly lapped over the inland Northwest. Bretz was vindicated, and in 1979 he was given the Penrose Medal, the geology world's equivalent of the Nobel Prize.

J Harlen Bretz (left) was the first to propose that the scablands had been formed by catastrophic floods.

Wheatfields (bottom) cover hills of loess – a rich soil that developed from dust borne on winds blowing off the ice sheets.

A talus pile (opposite) in the Palouse River Canyon is the product of erosion after the great floods.

Small-Scale Views

Dry Falls is certainly the greatest spectacle in the Channeled Scablands, but it's geology on the macro level. For a more intimate look at the scablands phenomenon, continue down Highway 17 south from **Soap Lake**. Here the road crosses the boulder-strewn Ephrata fan, a spray of debris left by the floods. *Pathfinder* scientists studied it as an analogue for Ares Vallis, the spacecraft's Martian landing site.

Then return to Soap Lake and head east toward Odessa en route back to Spokane. This town of about 1,000 people is on state Highway 28. Stop at the kiosk along First Avenue for maps to the scablands. Many prominent formations lie along a short loop drive north of town on state Highway 21 and Lakeview Ranch Road. There's spiral-shaped **Cinnamon Roll Crater**, tipped on a hillside; **Cache Crater**, a 50-foot hole that Native Americans may have used for food storage; and **Lakeview Ranch Crater**, with a center mesa the size of four football fields. Geologists theorize that these odd formations were created when lava flows erupted some 14 million years ago, leaving craters in their wake. The dike-rimmed craters were then sealed with lava, only to be exhumed by

shocked if their conceptions of an orderly world are challenged," he wrote in 1928. Standing on the coulee rim, hearing only the echoes of calling birds, it's virtually impossible to imagine the roar that must have accompanied Bretz's magnificent floods.

the Ice Age floods millions of years later.

The town of Odessa teamed with the Bureau of Land Management to create a nine-mile hiking, bicycling, and horseback-riding trail from the north edge of town to Lakeview Ranch. Along the way, visitors may spy deer and coyote or see golden eagles and red-tailed hawks aloft overhead. The Bureau's Lakeview Ranch area offers camping alongside Pacific Lake and opportunities for boating, fishing, and hunting in season.

Most of the scablands region is privately owned, but some landowners welcome visitors. At **Paradise Llama Ranch**, 11 miles northeast of Odessa, Tom and Debbie Weishaar hitch up Jerry and Katie, their Belgian mules, for covered-wagon treks to such unusual features as a natural bridge big enough to support a semi-truck. The ranch is especially pleasant in spring, when buttercups and bluebells burst forth, and in fall, when summer's 100°F-plus temperatures finally lose their grip on this arid land. Mountain biking opportunities also abound, but many visitors are content just to savor the silence and scenery.

A loop trip to Grand Coulee, Dry Falls, and Odessa is possible in one long day. If you have more time, consider heading south to one of the scablands' other grand sights at **Palouse Falls State Park**, about 75 miles southeast of Odessa. Dropping nearly 200 feet into a circular pool, Palouse Falls' ribbon-like cascade appears unusually delicate – until you realize that, once again, the scale of the landscape has tricked you. This area was forever changed when the Ice Age floods hurtled through, tearing away more than 300 feet of soil and transforming the **Palouse River** channel into a canyon 1,500 feet wide and 300 feet deep.

From Grand Coulee to Palouse Falls, the remoteness, stillness, and mystery of scablands country combine to make it a place far removed from the mainstream – a place where it's possible both to relax and to ponder geological forces on a grand scale. Landscapes such as the Channeled Scablands free the mind to wander and wonder, to dream big dreams and challenge the boundaries of what we know. Like J Harlen Bretz long ago and NASA's *Pathfinder* scientists today, we, too, can be geological visionaries.

TRAVEL TIPS

DETAILS

When to Go

The weather in eastern Washington is most pleasant in spring and fall. Daytime highs are in the low 60s in April, in the low 70s in September. Summer highs soar into the 90s and low 100s. Rain is sparse, but snowfall is fairly common October through March. The Grand Coulee area is heavily visited in summer.

How to Get There

Commercial airlines serve Spokane International Airport in eastern Washington. Amtrak stops in Spokane and Ephrata.

Getting Around

Driving is the only convenient way to explore the Scablands. Rentals are available at the airports.

Handicapped Access

Visitor facilities and numerous walking trails at Grand Coulee Dam are wheelchair-accessible, as is the nine-mile trail from Odessa to Lakeview Ranch.

INFORMATION

Bureau of Land Management

Spokane District, 1103 North Fancher Road, Spokane, WA 99212; tel: 509-536-1200.

Grand Coulee Dam Chamber of Commerce

P.O. Box 760, Grand Coulee Dam, WA 99133; tel: 800-268-5332 or 509-633-3074.

Odessa Chamber of Commerce

3 West First Avenue, P.O. Box 458, Odessa, WA 99159; tel: 509-982-0049.

CAMPING

Lakeview Ranch Recreation Area

Bureau of Land Management, 1103 North Fancher Road, Spokane, WA 99212; tel: 509-536-1200.

Primitive campsites are available at Lakeview Ranch, north of Odessa off Highway 21. The area has picnic tables, fire pits, outhouses, and a boat dock.

Odessa Golf Club

P.O. Box 621, Odessa, WA 99159; tel: 509-982-0093.

There are six hookups for recreational vehicles at this municipal golf course on the west end of town on Highway 28.

Spring Canyon Campground

National Park Service, 1008 Crest Drive, Coulee Dam, WA 99116; tel: 509-633-9188.

Eighty-seven sites are allotted on a first-come, first-served basis at this campground, part of the Lake Roosevelt National Recreation Area.

Steamboat Rock State Park

P.O. Box 370, Electric City, WA 99123; tel: 509-633-1304.

Located 12 miles south of Grand Coulee Dam on Highway 155, this campground is open year-round and has 26 standard campsites, 100 sites with full hookups, and 92 primitive sites.

Sun Lakes State Park

34875 Park Lake Road NE, Coulee City, WA 99115; tel: 509-632-5583.

This year-round park provides 174 campsites and 18 hookups.

LODGING

PRICE GUIDE – double occupancy

$ = up to $49	$$ = $50–$99
$$$ = $100–$149	$$$$ = $150+

Columbia River Inn

10 Lincoln Avenue, Coulee Dam, WA 99116; tel: 800-633-6421 or 509-633-2100.

This popular, 34-room motel is set on a hill just across the street from Grand Coulee Dam. Some rooms have balconies with views of the dam's nightly laser light show. A hot tub and outdoor pool are on the premises. Walking and biking trails are nearby. $–$$

Four Winds Guest House

301 Lincoln Avenue, Coulee Dam, WA 99116; tel: 800-786-3146 or 509-633-3146.

Four Winds was built in the 1930s as housing for engineers who worked on the dam, three blocks away. There are 10 guest rooms, most with shared baths. A gourmet breakfast is served each day, and a picnic and barbecue area is on the grounds. $$

Notaras Lodge

13 Canna Street, Soap Lake, WA 98851; tel: 509-246-0462.

This inn features hand-hewn log cabins decorated in a historical theme. Each of the 14 units has a kitchenette and mineral water piped in from Soap Lake. A 1,500-square-foot executive suite with full kitchen and Jacuzzi is also available. $–$$$

Odessa Motel

609 East First Avenue, Odessa, WA 99159; tel: 509-982-2412 or 877-414-9103.

This small, quiet motel on the east side of town offers 10 rooms and two apartments within walking distance of parks and restaurants. $–$$

TOURS

Grand Coulee Dam

P.O. Box 620, Grand Coulee, WA 99133; tel: 509-633-9265.

Guided tours of Grand Coulee Dam are offered daily at the Bureau of Reclamation's Visitor Arrival Center on Highway 155.

The route varies, but one highlight is usually the elevator ride into the third powerhouse, where the world's largest generators can be viewed. Narrated laser light shows are presented nightly June through September.

Lake Roosevelt National Recreation Area

1008 Crest Drive, Coulee Dam, WA 99116; tel: 509-633-9441.

The National Park Service offers guided canoe trips on Crescent Bay Lake in the Grand Coulee Dam area. The trips, usually held two or three times a week in the summer, include interpretation of the area's geology, wildlife, and history.

Paradise Llama Ranch

10463 Sommers Road East, Odessa, WA 99159; tel: 509-982-2404.

This outfit leads covered wagon trips into the scablands north of Odessa. Guests can also camp on the ranch and load up a llama or two for a pack trip into the surrounding area. Inquire about scenic flights over the scablands.

MUSEUMS

Dry Falls Interpretive Center
Sun Lakes State Park, 34875 Park Lake Road NE, Coulee City, WA 99115; tel: 509-632-5583.

This is the best place to learn more about the catastrophic floods that created the scablands. A short film tells the story of J Harlen Bretz, who first theorized that the odd terrain of central Washington was the result of epic floods. Outside, there's a staggering view of Dry Falls.

Excursions

Craters of the Moon National Monument
P.O. Box 29, Arco, ID 83213; tel: 208-527-3257.

Set amid the largest lava field in the Lower 48 states, Craters of the Moon served as a training site for lunar-bound NASA astronauts. This is the showplace of Idaho's Great Rift, a system of fissure vents, volcanic cones, and lava flows that erupted between 15,000 and 2,000 years ago. A seven-mile loop road leads to numerous trails to such intriguing formations as the Devil's Orchard, Spatter Cones, and Big Cinder Butte.

Columbia River Gorge National Scenic Area
902 Wasco Avenue, Suite 200, Hood River, OR 97031; tel: 541-386-2333.

The basalt flows of the Columbia River Plateau were laid down about 17 million years ago. Some of the most dramatic examples are in the Columbia River Gorge, which is characterized by five- or six-sided columnar jointing and rough, random-looking joints, often topped by "frozen" bubbles of lava. The Columbia Gorge Interpretive Center in Stevenson, Washington, has an excellent multimedia show on the area's geological history.

John Day Fossil Beds National Monument
HCR 82, Box 126, Kimberly, OR 97848; tel: 541-987-2333.

With three units spread across north-central Oregon, this 14,000-acre monument boasts an unusually complete and well-preserved record of plant and animal life from 45 million to 5 million years ago. The Clarno Formation is especially well known for its tropical and subtropical forest plant beds dating back 35 million years or more, while the John Day and Mascall formations have yielded an abundance of mammal fossils. The visitor center and museum are at the Sheep Rock Unit eight miles west of Dayville on U.S. Highway 26.

Glacier Bay
National Park
Alaska

CHAPTER 21

Visitors entering **Glacier Bay** from **Icy Strait** in midsummer are surrounded by abundance. Berry-eating black bears and grazing moose roam lush tidal meadows. Shellfish beds carpet the lower intertidal zone, which gives way to kelp forests patrolled by sea otters and seals. Flocks of gulls wheel over offshore tidal currents. All this is presided over by humpback whales and their occasional predators, the orcas. ◆ Yet there's a paradox here. *Glacier* Bay. *Icy* Strait. The austere names seem wildly out of step with the riot of life all around. The only ice in view lies in mountains 50 miles to the north. But it wasn't always so. ◆ Consider the same scene two centuries before: "We opened a large bay choakd up with ice and backd by a considerable tract of country presenting a prospect the most bleak and barren that can possibly be concievd ... not a tree or shrub could be percievd over this lifeless and drear tract, which with the surrounding ice diffusd a piercing chill we could scarcely endure." ◆ So wrote Archibald Menzies, who came here with Captain George Vancouver's expedition in July 1794. Nearly the whole of Glacier Bay was filled with ice then, extending to within 10 miles of the bay's mouth. The ice dominated the entire area with its enveloping presence, freezing breath, and legions of icebergs. ◆ Yet the ice had already been retreating for several decades when Menzies described the place. This withdrawal from what is now called the Little Ice Age continues right up to the present, leaving behind an array of fresh glacial landforms and opening a vast new territory to animal and plant recolonization.

Life slowly returns to a land laid bare by the retreat of glacial ice.

A kayaker in Glacier Bay's West Arm explores the face of Reid Glacier, one of a dozen tidewater glaciers that spills into the sea.

where. The forest that crowds along the roads and paths can't hide the chaos of landforms. Nor does the carpet of moss on the forest floor cover the boulders scattered everywhere. This is a glacial moraine – debris bulldozed by the ice snout into heaps of rubble. There's no sign of a previous generation of trees, and if you count the rings on stumps, you'll find that the trees there now are less than 200 years old. This makes perfect sense; when Menzies looked into the bay in 1794, ice was just leaving the area.

A walk on the beach fills out the story a bit more. Waves have sorted glacial rubble into bands that parallel the tide line. Beach grass and fireweed grow on two such bands. But above the reach of the tide, vegetation also lies in bands parallel to the shore, growing on older beach deposits. This shows that the land is rising, moving former beaches beyond water's reach and turning them over to colonizing waves of plants from the forest. The uplifting of the land – at a rate of about an inch a year – is mostly due to the recent removal of the weight of glacial ice.

The **Beardslee Islands**, about 10 miles off Gustavus, show that rivers, as well as the sea, have helped the glaciers sculpt the land. Paddle a kayak through the **Bartlett River** estuary into this maze of islands and examine the exposed sediments along wave-washed headlands. They are made of banded, sorted, and rounded sediments varying from fine-grained silt to coarse gravel – the signatures of running water. The Beardslees are the remains of a large glacial river plain. Radiocarbon dating of one buried tree stump shows that the plain was being built about 800 years ago. It was then home to the Tlingit people, whose oral history tells of

To tourist and scientist alike, the theme of **Glacier Bay National Park and Preserve** is change. Miss the motion and you miss the plot of a planetary drama. Not surprisingly, the main character of this drama – at least in the last act – is ice.

Glacial Evidence

A good place to begin piecing together Glacier Bay's story is near **Glacier Bay Lodge** at park headquarters, about 10 miles north of **Gustavus**. Lessons from the ice are every-

The Beardslee Islands (right) are the remains of a glacial outwash floodplain, now half submerged by the sea.

Small chunks of ice (below) called "bergie bits" are scattered throughout the bay.

A boardwalk (opposite) winds through a spruce and hemlock rain forest at Bartlett Cove; glaciers retreated from the area about 200 years ago.

eviction by ice. Glacial rubble on top of the river sediments corroborates their story.

Exploring Glacier Bay by tour boat is like going back in time. The glacier that occupied the bay in Menzies' day has been shrinking northward, uncovering one spot after another along the shores. Whereas the area around park headquarters is about 200 years old and supports a tall spruce forest, the shores along **Whidbey Passage** are about 60 years younger. The forest is just reaching the spruce stage here. But at **Sandy Cove** or **Marble Mountain**, gaze up about 2,000 feet and you will see a "bathtub ring" of ancient forest, too high for the glacier to have touched.

Traces of lowland forests linger, too, if you look closely. The shores of the mid-bay islands bristle in places with stumps exhumed by waves and stained gray by silt and minerals. Radiocarbon dating

shows that they died several thousand years ago. These trees were growing during an interglacial time, prior to the Little Ice Age. So what killed them?

The answer lies below a veneer of glacial rubble, where the stumps' roots are covered by deposits like those that make up the Beardslee Islands. As the glacier came down the bay, it built a river plain in front of it. The glacier's outwash dumped sediments onto the plain, and the debris overwhelmed the trees. Then the ice rumbled forward and tore the forest to shreds.

Uphill along the leading edge of the glacier is an area of rubble laid out in a rectangular pattern. This is the remains of material that fell into glacial crevasses when the ice was here a few decades ago. Somewhat larger are the eskers, sinuous ridges of sediments carried by streams flowing in and under the stagnant ice. A stream near the glacier's left margin cuts into glacial deposits containing wood dating from about 4,600 years ago. This marks the onset of the ice advance that culminated at Icy Strait, just before Menzies' visit.

Shoals tailing off either side of the inlet's mouth are moraines abandoned by **Reid Glacier** a little after 1900. On the west wall, a set of terraces begins behind the moraines and gains elevation up-inlet, ending well above the glacier. Walking along the terraces, you find that they were built by small streams and ice action along the glacier's margin when the ice extended to the inlet mouth.

Living Ice

Continuing northwest up the **West Arm** by tour boat takes you into an increasingly spectacular fjord-scape delimited by glacier-scoured cliffs and peaks. Side fjords and hanging, U-shaped valleys intersect the main channel, giving views up into the mountain-ous, snow-clad hinterlands. The young forests of the bay give way to even younger shrub thickets. All signs point to glaciers just ahead.

As the boat pulls past the gleaming white marble of **Ibach Point**, **Reid Inlet** suddenly opens to the left, a U-shaped, tundra-clad cleft between high granite ridges. At its head, you finally see the living ice, spilling out of the **Brady Icefield** and descending abruptly to the shore.

Glaciers Everywhere

Rounding **Jaw Point**, you are suddenly in a deep fjord that cleaves straight into the heart of the **Fairweather Range**. Along this grand-est of all Glacier Bay inlets, peaks and spires rise directly from sea level to nearly 9,000 feet. Glaciers are everywhere you look, cascading down side valleys to the water's edge, adher-ing to headwalls, or plunging over cliffs.

At the fjord's head, **Johns Hopkins Glacier** writhes out of the canyon it has carved amid even higher mountains. Its surface is nearly black with debris it has ripped from the landscape. The Johns Hopkins sheds more icebergs than any other glacier in the park, creating an icepack fastness inhabited by thousands of harbor seals during pupping and molting seasons. There are virtually no glacial deposits, except deep below, on the fjord bottom. Here erosion is king.

Over thousands of years, glacial ice carved deeply into the Fairweather landscape, while forces of plate tectonics pushed the mountains ever higher. Geologists calculate that in the last 25 million years or so, these mountains have risen about eight miles. As they rose, erosion tried to tear them down. The fact that **Mount Fairweather** stands 15,000 feet high shows that erosion lost this particular race by nearly three vertical miles. But in **Johns Hopkins Inlet**, the ice has managed to saw down through all eight miles of rock.

Like a monster from the past, Johns Hopkins Glacier lies amid this magnificent desolation. Weakened by the warm modern climate, it waits for the day when it can creep down the inlet and reoccupy the great bay it helped to create.

Lamplugh Glacier (opposite) drops from the Fairweather Range into the bay. The dark stripes are moraines – rivers of rocky debris fed by tributary glaciers.

Dwarf fireweed (top right) is one of the first plants to resettle on ground newly uncovered by glaciers.

The edge of Reid Glacier (below) exhibits the unmistakable blue glow of glacial ice.

Glaciers and Climate Change

Glaciers are made of ice, a substance that scarcely survives at the temperatures common on the Earth's surface. This makes glaciers very sensitive indicators of climatic change.

Glacial ice has only rarely been prevalent during geologic time. One such period is the Pleistocene, beginning two million years ago, when ice repeatedly crept out of the mountains and high latitudes to cover as much as a fifth of the world's landmasses, only to shrink before the onslaught of warmer times.

The causes of glacial cycles are not fully understood, but scientists suspect that regular changes in the Earth's orbit around the Sun and the wobble in its rotation seem to lie at the root. These changes influence the amount of solar heating around the globe. Variations in the quantity of dust and the proportions of gases in the atmosphere also play a major role. These determine how much incoming solar heat is absorbed by the ground or sent back into space.

Gases trapped in the ice sheets covering Greenland and Antarctica are a chronicle of climate, atmospheric composition, and glacial history extending thousands of years into the past. That record shows a strong correlation between the amount of carbon dioxide in the atmosphere and the global temperature.

Like a greenhouse window, carbon dioxide allows light through, but once sunlight changes into heat, the gas prevents it from escaping, and the world grows warmer. Concentrations of carbon dioxide are as high as they have been for at least a thousand years. And sure enough, temperatures appear to be rising on a global scale. This is tipping the thermal balance against glaciers, and most of the world's ice is indeed shrinking – in some cases, at an accelerating pace.

TRAVEL TIPS

DETAILS

When to Go

The park is open year-round, the visitor center from mid-May to mid-September. Visitation peaks in July. Average July temperatures range from 48° to 63°F; January, 26° to 16°F. Rain and cold are possible in any season, so pack a hat, gloves, rain gear, and waterproof boots.

How to Get There

Commercial airlines serve Juneau International Airport, 70 miles from Glacier Bay. There is no road to the park, which is reached only by charter boat and plane. Scheduled air service is available from Juneau to Gustavus; buses and taxis run the 10 miles between Gustavus and the park entrance. Contact the park office for a list of transportation services.

Getting Around

Kayaking is popular with independent travelers in the park. Kayak rentals and guides are available. The majority of travelers view Glacier Bay from cruise ships. Daily cruises depart from Glacier Bay Lodge, nine miles from Gustavus. Contact the park for details on boating permits and restrictions.

Backcountry Travel

Hiking and backpacking are allowed in the park; no fee or permit is required, but all backcountry travelers are advised to contact rangers before departing.

INFORMATION

Alaska Public Lands Information Center
605 West 4th Avenue, Suite

105, Anchorage, AK 99501; tel: 907-271-2737 or 3031 Tongass Avenue, Ketchikan, AK 99901; tel: 907-228-6220.

Glacier Bay National Park
P.O. Box 140, Gustavus, AK 99826-0140; tel: 907-697-2230.

Gustavus Visitor Association
P.O. Box 167, Gustavus, AK 99826; tel: 907-697-2475.

CAMPING

The park campground at Bartlett Cove is available on a first-come, first-served basis. Campers must attend a free camper orientation at the Visitor Information Station and check out a bear-resistant food canister. Bear-resistant food caches, firewood, and a warming hut are provided at the campground.

LODGING

PRICE GUIDE – double occupancy	
$ = up to $49	$$ = $50–$99
$$$ = $100–$149	$$$$ = $150+

Bear Track Inn
P.O. Box 255, Gustavus, AK 99826; tel: 888-697-2284 or 907-697-3017.

This handcrafted log lodge has 14 spacious guest rooms with private baths, two queen-sized beds, and down comforters. The lobby, whose ceiling rises 30 feet, has a walk-around fireplace, overstuffed suede couches, and moose-antler chandeliers. Packages include gourmet meals, ferry transportation from Juneau, and all ground transportation. Guests choose from activities such as whale watching, sea kayaking, and airplane tours. Open February 15 to October 1. $$$$

Glacier Bay Lodge
Glacier Bay Park Concessions, 520 Pike Street, Suite 1400, Seattle, WA 98101; tel: 800-451-5952 or 206-626-7110.

Striking views of the rain forest are available from the park's only lodge, set on Bartlett Cove, nine miles from the airport in Gustavus. More than 50 rooms, each with three bunk beds, are available in men's and women's dormitories. Guests share bath and shower facilities. Breakfast, lunch, and dinner are served in the dining room; patio dining is available in the evening. Boat tours, which leave from the dock below the lodge, and a camper drop-off service are available. Open May 15 to September 10. $

Gustavus Inn
P.O. Box 60, Gustavus, AK 99826 (May to September) or 7920 Outlook, Prairie Village, KS 66208 (October to April); tel: 800-649-5220 or 907-697-2254.

This cedar-sided, New England-style house was built in 1925 and later expanded. Overlooking Icy Strait about eight miles from the park, the inn offers 13 rooms, each with a private bath, a queen-sized bed, and one twin bed. Included in the price are three daily meals, featuring fresh vegetable and seafood dishes. Boat tours, fishing charters, and sea kayaking are offered. Open May 15 to September 15. $$$

Meadow's Glacier Bay Guest House
P.O. Box 93, Gustavus, AK 99826; tel: 907-697-2348.

Guests enjoy gourmet breakfasts, the use of bicycles, fishing poles, and binoculars, and free transportation to and from the dock or airport. Tours can be arranged from the guest house, including a kayaking trip with a naturalist. $$–$$$

Whalesong Lodge
P.O. Box 389, Gustavus, AK 99826; tel: 800-628-0912.

Accommodations range from guest rooms with private baths to condominiums. Arrangements can be made for overnight tours, private day charters, guided kayaking trips, and whale-watching excursions. $$–$$$

TOURS

Alaska Discovery Wilderness Adventures

5449 Shaune Drive, Juneau, AK 99801; 800-586-1911 or 907-697-2411.

Guides, equipment, and instruction are available for sea kayaking, rafting, canoeing, and camping adventures. The company also runs a bed-and-breakfast in Gustavus, and offers inn-to-inn wilderness trips.

Alaska Mountain Guides and Climbing School

P.O. Box 1081, Haines, AK 99827; tel: 800-766-3396 or 907-766-3366.

The school offers a guide service, climbing instruction, and outfits expeditions. Packages include a Glacier Bay ski traverse, a glacier hike with ski plane, and a 10-day trip combining glacier trekking with sea kayaking.

Glacier Bay Sea Kayaks

P.O. Box 26, Gustavus, AK 99826; tel: 907-697-2257.

The park concession offers sea-kayak rentals, outfitting, orientation and instruction, and can arrange drop-off and pickup service on the park tour boat.

Northgate Tours and Cruises

P.O. Box 20613, Juneau, AK 99802; tel: 888-463-5321 or 907-463-5321.

Three- to five-night "active adventure" cruises aboard the 86-passenger M/V *Wilderness Discoverer* explore Glacier Bay and the Inside Passage. Passengers sleep on board and spend each day sea kayaking or hiking on shore.

Wild Alaska Glacier Bay Adventure

P.O. Box 335, Gustavus, AK 99826; tel: 800-225-0748.

Package tours include a cruise through the Inside Passage's steep mountains and deep fjords and whale-watching excursions. Custom itineraries can be arranged.

Excursions

Kenai Fjords National Park

P.O. Box 1727, Seward, AK 99664; tel: 907-224-3175.

The Kenai Mountains are being dragged slowly under the sea by the collision of two tectonic plates. Alpine valleys once filled with glacier ice are now deep-water fjords flanked by jagged peaks. In 1964, an earthquake dropped the shoreline six feet in just one day. As the land sinks into the ocean, glacier-carved cirques are turned into half-moon bays and mountain tops are reduced to wave-beaten islands. Visitors can stroll a half-mile to the crevassed face of Exit Glacier and see newly exposed, scoured bedrock.

Wrangell-St. Elias National Park and Preserve

P.O. Box 439, Copper Center, AK 99573; tel: 907-822-5234.

Glaciers and icefields cover nearly a third of this wilderness park, which has nine of the 16 highest peaks in the United States. Malaspina Glacier, fed by 25 tributary glaciers and larger than Rhode Island, spills onto the forelands below Mount St. Elias. Opportunities for exploring include river rafting, flightseeing, and kayaking in Icy Bay, where tidewater glaciers calve icebergs into the sea. Glacial retreat opened up the bay, now 25 miles long, starting in the early 1900s.

Denali National Park and Preserve

P.O. Box 9, Denali Park, AK 99755-0009; tel: 907-683-2294.

Mount McKinley, rising 20,320 feet, is the granite-and-slate centerpiece of this park in the Alaska Range, which began to rise some 65 million years ago along the Denali Fault, North America's largest crustal break. As they rose, the mountains were eroded, sculpted, and weighed down by huge masses of ice. Numerous glaciers still radiate from the high peaks. Access is carefully regulated; most visitors explore along a 90-mile gravel road that runs through the park. The debris-laden snout of 35-mile-long Muldrow Glacier lies within a half mile of the park road.

Hawaii Volcanoes
National Park
Hawaii

CHAPTER 22

O n January 3, 1983, **Kilauea** awoke. ◆ Swarms of earthquakes shook the island of Hawaii. The ground heaved and ripped open as magma forced its way to the surface. A throbbing roar broke the silence, and fiery spasms of lava shot skyward. Stinging sulfurous fumes saturated the air as gases exploded, hurling flaming red clots of molten rock across the desolate landscape. The air felt like the breath of a blast furnace. Then, after three weeks, all activity ceased, only to flare up again with similar fury the following month. ◆ Thus began what is now the longest recorded eruption at **Hawaii Volcanoes National Park**. As of late 1999, Kilauea was continuing to ooze lava from its craggy surface. ◆ The Hawaiian islands blazed to life millions of years ago as magma erupted through a crack in the ocean floor. The crack **Periodic eruptions on the** marked a "hot spot," a place where magma rises **Big Island of Hawaii** from Earth's mantle and punches through **send rivers of molten rock** the overlying crustal plate. This hot spot created **oozing toward the sea.** a volcano on the Pacific seafloor, piling up countless lava flows to make what geologists call a shield volcano, a name that reflects its broad, gently domed shape. The hot spot was persistent and in time the shield volcano broke into the sunlight and became an island. ◆ For millions of years, the Pacific crustal plate has been creeping a few inches per year toward the northwest across the stationary hot spot. Off this geological assembly line, one by one, came a chain of volcanic islands. These ancestral Hawaiian islands, now called the Emperor seamounts, dot the ocean floor northwest of present-day Hawaii right up to the Aleutians in Alaska. As plate tectonics carried the islands away from their magma source, they stopped growing.

Pele, the Hawaiian goddess of fire,
speaks from the heart of Kilauea
volcano, illuminating the night
with a fountain of molten basalt.

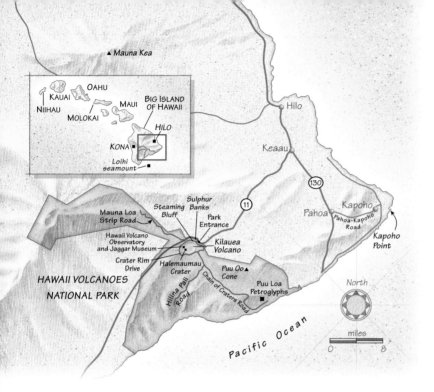

last eruptions occurred about 3,500 years ago. **Hualalai**, in the west, last erupted about 200 years ago, while **Kohala**, in the north, hasn't been active for 100,000 years. **Mauna Loa** and **Kilauea** are far more active.

Mauna Loa, meaning Long Mountain, is enormous. Its summit rises 13,677 feet above sea level; from the ocean floor it measures 31,000 feet, making it more than a thousand feet taller than Mount Everest. The volcano has erupted 18 times in the last century, and a major eruption could happen tomorrow – or 20 years from now. Since this unpredictability makes studying the volcano difficult, volcano watchers, amateur and professional, gravitate to Mauna Loa's little cousin, Kilauea. Its summit reaches only 4,100 feet above sea level, but it has been in eruption steadily since 1983 and erupted frequently before then.

Pele's Crater

Like Mauna Loa, Kilauea is situated within Hawaii Volcanoes National Park. The park entrance lies off Hawaii Route 11, about an hour's trip from **Hilo** or two hours from **Kona**. Stop at the visitor center for information and brochures, and then follow the 11-mile **Crater Rim Drive**. This paved road circles Kilauea's summit caldera, the broad pit formed as lava under the summit retreated and the rock layers collapsed. The drive passes through both lush rain forest and a lava desert.

A mile from the park entrance, your nose will detect **Sulphur Banks**, where piles of crystal-encased rubble smolder. A charred rotten-egg smell pervades the area. The crystals grow together like delicate yellow snowflakes and are best appreciated up close despite the stench. (Visitors with coronary or respiratory problems shouldn't

Swordfern (left) is one of the first plants to colonize fresh lava flows.

Offerings to Pele (opposite, top) lie on the rim of Halemaumau crater; the goddess is reputed to have a taste for gin.

A billowing column of steam (opposite, bottom) marks the spot where lava pours into the sea.

Wind, rain, and waves steadily attacked them until each sank from sight, back into the ocean from whence they had arisen.

Today's Hawaiian islands will eventually face a similar fate. First to go will be **Niihau**, then the islands of **Kauai**, **Oahu**, **Molokai**, and **Maui** and their neighbors, and finally the **Big Island** of Hawaii, the youngest and largest. Already, the next island, **Loihi**, is growing on the seafloor off Hawaii's southeast coast. For now, however, the hot spot is feeding magma mostly to the southeastern part of the Big Island.

All the island's peaks are volcanic. **Mauna Kea** is the highest (13,796 feet); its linger – the fumes can be hazardous.)

As you continue along the drive, the forest fringes off to a golden prairie shimmering with sedge grass. Columns of vapor float through a grassland known as **Steaming Bluff**. Steam forms when rain seeps through cracks to reach rocks heated by magma. At times, rainwater boils and hisses into towering shafts of vapor. A five-minute walk to the cliff's edge opens a majestic view of **Halemaumau Crater** at the summit of Kilauea. Trade winds carry the steam from the bluff down into the deep black pit in veils of cascading vapor.

The road loops around the outer edge of the 500-foot-deep caldera. The **Jaggar Museum** sits on the western rim at the **Hawaiian Volcano Observatory**, which was founded in 1912 and is now run by the U.S. Geological Survey. Seismographs in the museum record hundreds of earthquakes beneath the caldera every day. Step outside to the precipitous edge for another panoramic view of the three-mile-wide caldera and

Halemaumau Crater. You'll see historic lava flows, pit craters, cinder cones, and giant cracks that could swallow whole tour buses.

A short drive from the museum curves down into Halemaumau Crater itself. The first Hawaiians 1,500 years ago watched in awe as Kilauea's bright red lava seemed to

and an exhilarating 3,700-foot descent to the coast. A series of enormous pit craters form a volcanic necklace along the upper part of the road. Pit craters occur as the surface collapses when underlying magma drains away.

Farther along this road are dramatic black lava cliffs called pali, lava sea arches, frozen "waterfalls" of lava draping the green cliffs, and one of the richest archaeological finds in Hawaii, the **Puu Loa petroglyphs**, ancient drawings carved into lava stones. This, too, is a sacred area, so magical that Hawaiians say it can give you "chicken skin," or goose bumps as mainlanders would put it.

scald the heavens, and they followed the glow to this pit. They believed the eruptions were the magic of an ancestral spirit-goddess. Her name is Pele, and she still lives in the hearts and souls of Hawaii's people as the ultimate life force. Here, at Halemaumau Crater, is where Pele resides, and Hawaiians expect visitors to respect her home.

Rock on the Move

After you leave the crater, a left turn leads to the 20-mile-long **Chain of Craters Road**

Don't be surprised when the road dead-ends abruptly. In 1995 a lava flow covered the highway and has now cooled. Kilauea has two types of lava: pahoehoe (pronounced PAH-hoy-hoy) and aa (ah-ah). Pahoehoe appears to have a smooth surface, is very fluid, and pools in rippled toes or races along in rivers. A'a is sharp, chunky, and rough, and can cut unprotected feet and hands.

At the end of Chain of Craters Road, an

Birth of an Island

While Kilauea entertains visitors with fireworks, Hawaii's youngest submarine volcano is emerging from the ocean floor 20 miles to the south in 3,200 feet of water. In a few tens of thousands of years, **Loihi** seamount will become the newest Hawaiian island.

Like Mauna Loa and Kilauea volcanoes, Loihi is forming a shield. The young volcano has frequent earthquakes, is dappled with steam vents spitting hot gases into the water, and experiences landslides. A caldera 1.7 miles wide and 2.3 miles long dominates Loihi's summit. One side is pocked with three large pit craters. The biggest, **Pele's Pit**, opened when a spasm of vigorous earthquakes jolted the seamount in July and August, 1996.

In an attempt to witness the volcano's undersea eruptions, marine geologists and geophysicists from the University of Hawaii climbed aboard a small submersible and bravely headed for the bottom. As they hovered in their tiny vessel, the ocean water around them was shaken by tremors deep in the volcano. As it happened, the scientists missed seeing an eruption in progress, but they left instruments behind to monitor Loihi and conduct experiments. An underwater microphone relays sounds from the volcano. The first sounds picked up, however, were the songs of migrating humpback whales.

A scuba diver (above) investigates underwater lava flows similar to those of the Loihi seamount.

Lava (opposite, top) pours into the sea.

The skeletons of Ohia trees (bottom) killed in the 1959 eruption of Kilauea are scattered about Devastation Trail.

underground lava tube, running seven miles from the active **Puu Oo Cone**, channels intermittent flows of lava over the craggy black cliffs into the Pacific. Lava from Puu Oo's eruptions has covered many square miles and in 1990 destroyed the town of Kalapana near the park.

To see the eastern side of this extensive flow, drive back toward Hilo on Route 11, then at Keaau turn right on Route 130. This road takes you across lava flows and past cinder cones several hundred years old as well as ones dating from 1955 and 1960. At the dead end on the eastern side of the big flow, the coast road turns and runs eastward, passing though the little towns of **Pahoa** and **Kapoho** perched on the flows that mark Hawaii's southeastern tip.

Whenever Kilauea sends lava down its flanks, visitors to the park can witness the birth of the newest land on Earth. For the best view of an active lava flow, arrive at the Puu Oo lava tube just before dusk. As the sun sets, the sky turns deep blue to match the sea. A soft orange glow brightens at the base of a sizable steam cloud. Hot lava drips over the sea cliffs into the ocean, instantly turning the water into clouds of vapor. Plan to watch for an hour or so as the stars blink on one by one. Often, as the waves retreat and ocean breezes momentarily clear the steam, you'll see a telltale strip of bright orange lava entering the sea against the black cliffs.

TRAVEL TIPS

DETAILS

When to Go

Hawaii enjoys subtropical weather with moderate humidity year-round, but climate varies considerably within the park. Upland rain forests receive almost 200 inches of rain a year, and the average temperature is 65°F. The coast is warmer and drier, with average temperatures hovering around 75°F. November through April is the rainy season, with 10 to 13 inches of precipitation per month, depending on location, and temperatures in the 50s and 60s. Expect temperatures in the 70s May through October.

How to Get There

Major airlines and interisland carriers serve Keahole Airport in Kona, 100 miles from the park. Interisland carriers also serve General Lyman Field in Hilo, about 30 miles from the park.

Getting Around

An automobile is necessary for touring the park. Car rentals are available at the airports.

Handicapped Access

The visitor center and Devastation Trail are accessible.

INFORMATION

Hawaii Volcanoes National Park
P.O. Box 52, Hawaii Volcanoes National Park, HI 96718; tel: 808-985-6000.

Big Island Visitors Bureau
250 Keawe Street, Hilo, HI 96720; tel: 808-961-5797.

CAMPING

Two park campgrounds, Namakini Paio and Kulanaokuaiki, have sites available on a first-come, first-served basis. Namakini Paio has 10 cabins with electricity and barbecue pits. To reserve a cabin, call 808-967-7321.

Backcountry Travel

A permit, obtainable one day in advance, is required for camping and backcountry travel in the park. For information, call 808-985-6000.

LODGING

PRICE GUIDE – double occupancy

$ = up to $49 $$ = $50–$99
$$$ = $100–$149 $$$$ = $150+

All Islands Bed-and-Breakfast
463 Iliwahi Loop, Kailua, HI 96734; tel: 800-542-0344 or 808-263-2342.

The agency arranges lodging at more than 700 bed-and-breakfasts throughout the Hawaiian Islands and assists with other travel plans. Prices vary.

Arnott's Lodge and Hiking Adventures
98 Apapane Road, Hilo, HI; tel: 808-969-7097.

This 25-room lodge offers Internet service, a free shuttle, expeditions, and bicycle rentals and allows people to set up tents on its lawn for a small fee. Accommodations range from bunks with shared bathroom facilities to a private suite. $–$$$

Chalet Kilauea Collection
P.O. Box 998, Volcano, HI 96785; tel: 808-967-7786.

The service represents about 30 inns and rental houses, ranging from economical bed-and-breakfasts to spacious vacation homes. Prices vary.

Country Goose Bed-and-Breakfast
P.O. Box 597, Volcano, HI 96785; tel: 800-238-7101 or 808-967-7759.

Perched at an elevation of 3,500 feet, the Country Goose offers two guest rooms, each with a private bath and entrance. Gourmet breakfasts are served. $$

Hale Ohia Cottages
P.O. Box 758, Volcano, HI 96785; tel: 800-455-3803 or 808-967-7986.

Three cottages with kitchen facilities occupy a one-acre garden filled with native plants. Four suites have private baths and refrigerators; a deluxe suite has a separate living room and bedroom. Breakfast is included. A hot tub is available for use in a private corner of the garden. $$–$$$

Horizon Guest House
Box 268, Honaunau, HI 90726; tel: 808-328-2540.

Set on 40 acres on Mauna Loa, this guest house offers four bedrooms, ocean views, and eco-tours. The lodge is wheelchair-accessible. $$$$

Kalahiki Cottage
McCandless Ranch, P.O. Box 500, Honaunau, HI 96726; tel: 808-328-8246.

Situated on a 15,000-acre ranch that stretches from the west coast of the Big Island to the slopes of Mauna Loa, this one-bedroom cottage has a kitchen and private lanai. The property is set in one of the largest intact forests on the island. $$$$

Kilauea Lodge
P.O. Box 116, Volcano, HI 96785; tel: 808-967-7366.

This 12-room lodge is set on 10 acres about a mile from the park. A two-bedroom cottage has a kitchen and living area. A restaurant is on the premises. $$$–$$$$

Pineapple Park Hostels

P.O. Box 639, Kurtistown, HI 96760; tel: 808-968-8170, or 1-877-865-2266.

Accommodations at this secluded inn range from inexpensive bunks to private bungalows. Mountain bike rentals and tours of the volcano and lava tubes are available. $–$$$

Volcano House

P.O. Box 53, Hawaii Volcanoes National Park, HI 96718; tel: 808-967-7321.

Volcano House has more than 42 units near the rim of Kilauea caldera, ranging from small accommodations to deluxe second-floor rooms with views of the crater. A restaurant is on the premises. $$–$$$$

TOURS

Hawaii Air Tours

200 Kanoelehua Avenue, Suite 103-285, Hilo, HI 96720; tel: 877-228-5954.

Helicopters and small airplanes depart from Hilo and Kona for flightseeing tours of active volcanoes, Hilo's waterfalls, and Kalapana, a village destroyed by lava. The company also coordinates hiking tours, including forays to Kilauea.

Hawaii Volcanoes National Park

P.O. Box 52, Hawaii Volcanoes National Park, HI 96718; tel: 808-985-6000.

Park rangers and naturalists lead several nature walks daily.

Excursions

Waimea Canyon

Koke'e Natural History Museum, P.O. Box 100, Kekaha, Kauai, HI 96752; tel: 808-335-9975

Mount Waialeale dominates the island of Kauai and is one of the rainiest places on Earth. Torrents pouring down the slopes have eroded deep canyons. The largest, Waimea, known as the Grand Canyon of the Pacific, is 10 miles long and nearly 3,000 feet deep. Stop at the Koke'e Natural History Museum for trail maps of Waimea Canyon State Park and neighboring Kokee State Park. Guided hikes are available June through September. The 1.8-mile Canyon Trail follows the rim of Waimea and offers views of both Waimea and Poomau canyons. The Kukei Trail is a scenic but steep trek into the canyon.

Diamond Head

Oahu Visitors Bureau, 2270 Kalakaua Avenue, Honolulu, HI 96815; tel: 877-525-6248 or 808-923-1811.

This extinct volcanic ash cone, rising 761 feet above Waikiki beach, was given its modern name by early explorers, who found calcite crystals here and mistook them for diamonds. (The calcite came from fragments of coral reef that were dissolved and re-precipitated in groundwater heated by the volcano.) Hawaiians originally named the volcano Leahi and believed that Pele, the fire goddess, lived there until she moved to the Big Island. A hiking trail leads up the west wall of the crater. As you climb, you'll see concrete fortifications built during World War II, when the military stored weapons here. The view from the top is spectacular.

Haleakala National Park

P.O. Box 369, Makawao, HI 96768; tel: 808-572-9306

Haleakala volcano last erupted in 1790, leaving behind an otherworldly landscape of sheer cliffs, shimmering waterfalls, jumbled lava flows, and multicolored cinder cones. Visitors can explore more than 30 miles of hiking trails, some leading into the crater valley, more than two miles wide and half a mile deep. The valley formed when the heads of two eroding stream valleys converged near the summit; it was later reshaped by lava flows and cinder-cone eruptions.

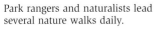

Resource Directory

FURTHER READING

General Reference

Earth science reference books fill vast libraries, but the basics you'll want to take along on a field trip can fit into a space not much bigger than a six-pack. While most of these titles are available in new editions, you would do well to try used bookstores. Also, get to know the earth science collections at your local libraries, both public and university.

Annals of the Former World, by John McPhee (Farrar, Straus and Giroux, 1998). Winner of the Pulitzer Prize in 1999, this book presents all of McPhee's stories about geology and geologists in one volume. It's not essential in the field, but it makes a great traveling companion.

Dictionary of Geological Terms, 3rd ed., edited by Robert L. Bates and Julia A. Jackson (Anchor Press, 1984). This paperback can fit in a backpack. A newer, greatly expanded edition, *Glossary of Geology* (1997), is the size of a small unabridged dictionary.

Earth from Above, by Yann Arthus-Bertrand (Abrams, 1999). While this book leans more toward natural-history subjects as seen from a low-flying airplane, there is much here for anyone interested in the geosciences.

Earth from Above: Using Color-Coded Satellite Images to Examine the Global Environment, by Claire L. Parkinson (University Science Books, 1997). The book delivers exactly what the subtitle promises. The expected audience is the interested lay reader and beginning university students.

Earth's Dynamic Systems, by W. Kenneth Hamblin (Macmillan, 1989). This is easily the best guide to how geological forces and processes operate in the world around us. There are superb line drawings.

A Field Manual for the Amateur Geologist, by Alan M. Cvancara (John Wiley & Sons, 1995). This volume presents thorough, practical information on how to do geology. If you own just one book on the subject, this should be it.

Geology Illustrated, by John S. Shelton (W. H. Freeman, 1966). This classic guide to the formation of the North American landscape features loads of compelling aerial photographs.

The Home Planet, by Kevin W. Kelley (Addison-Wesley, 1988). This large-format collection of Earth photos taken from space has evocative text opposite the images and a more technical write-up of each image in an appendix.

Peace of Mind in Earthquake Country, by Peter I. Yanev (Chronicle Books, 1991). The book covers how earthquakes occur and what they do while in progress. It also provides a highly practical guide on how you can prepare to survive one with the least possible damage to person and property.

Principles of Geology, by Charles Lyell (University of Chicago Press, 1991). Darwin took the first two volumes of Lyell's book on the *Beagle* and the third volume, published later, caught up with him during the voyage. Lyell's notion that geological change occurs slowly over great spans of time was enormously influential when this book was published in the 1830s, and it remains a core tenet today.

Regional Geomorphology of the United States, by William D. Thornbury (John Wiley & Sons, 1965). This out-of-print and somewhat out-of-date volume is still a useful summary of the events that have shaped the various regions of the United States.

Volcanoes, by Robert and Barbara Decker (W. H. Freeman, 1997). This is the latest edition of a fine beginner's guide to volcanoes and how they work.

Children's Books

A variety of illustrated books will introduce young readers to basic geological concepts and field identification.

Incredible Earth, by Nick Clifford (Dorling Kindersley, 1996).

Earth Facts, by Cally Hall and Scarlett O'Hara (Dorling Kindersley, 1995).

The Kingfisher Young Peoples' Book of Planet Earth, by Martin Redfern (Kingfisher, 1999).

Geology, by Frank H. T. Rhodes (Golden Books, 1991).

Janice VanCleave's Earth Science for Every Kid, by Janice VanCleave (John Wiley & Sons, 1991).

The Story of Geology: Our Changing Earth Through the Ages, by Jerome Wyckoff (Golden Press, 1976).

Rock Identification Guides

Despite the prominence given to rocks in their titles, most of the available field guides are more helpful in identifying fine mineral specimens than the ordinary rocks you're likely to encounter outdoors. These are among the most helpful titles.

The Audubon Society Field Guide to North American Fossils, by Ida Thompson (Knopf, 1985). Geologists often encounter fossils in the field, and this illustrated handbook makes it easy to tell a trilobite from a crinoid.

The Audubon Society Field Guide to North American Rocks and Minerals, by Charles Chesterman (Knopf, 1979). The main focus is on minerals, but good photos of rocks, both as hand samples and outcrops in nature, make this a most useful guide for field identification.

A Field Guide to Rocks and Minerals (Peterson Field Guides), by Frederick H. Pough (Houghton Mifflin, 1998). This is an excellent guide and widely available, but unfortunately it deals rather summarily with rocks and is largely devoted to minerals.

Rocks & Minerals: An Explore Your World Handbook (Discovery Books, 1999). This current, pocket-sized guide pairs background information with crisp, color photos for easy field identification.

Rocks and Minerals, by Charles A. Sorrell (Golden Press, 1978). This book is concerned almost entirely with mineral specimens. Illustrations are artwork rather than photos, which is a drawback.

Rocks and Minerals, by Herbert S. Zim and Paul R. Shaffer (Golden Press, 1989). Pocket-sized, this Golden Guide covers mostly minerals, but devotes some space to rocks at the back. Illustrations are artwork rather than photos.

Rocks & Minerals (Macmillan Field Guides), by Pat Bell and David Wright (Collier/Macmillan, 1985). This book features photographs of rock samples (including several examples of each type), but doesn't have pictures of typical outcrops.

Regional Geology Guides

In addition to the titles listed below, readers should be aware of three fine series of regional geology guides organized by state. The first two, *Roadside Geology* and *Geology Underfoot*, published by Mountain Press, detail geology for nonspecialists. The third, *Rockhounding*, published by Falcon Publishing, covers sites for rock and mineral collectors.

Cape Cod and the Islands: The Geologic Story, by Robert N. Oldale (Parnassus Imprints, 1992). This book supersedes Arthur Strahler's 1966 *A Geologist's View of Cape Cod* and covers recent changes to the landscape and a more up-to-date scientific understanding. It also provides suggestions for a geological road trip.

Centennial Field Guide (Geological Society of America, 1986-87). Written by geologists for other geologists, these six volumes make no concession to beginners, but the field trips go to hundreds of interesting and significant geological sites. The volumes are organized around regions of North America: Northeast, South-Central, Rocky Mountains, and others. Look for the set in a university library and photocopy the sections of interest; field trips generally run four to six pages.

Colossal Cataract: The Geologic History of Niagara Falls, edited by Irving H. Tesmer (State University of New York Press, 1981). A collection of articles that covers all aspects of the falls. Written for the interested lay reader.

Cycles of Rock and Water, by Kenneth A. Brown (HarperCollins, 1993). This highly readable book explores the Pacific coast of North America, with an emphasis on what happens when two tectonic plates collide.

Death Valley: Geology, Ecology, Archaeology, by Charles B. Hunt (University of California Press, 1975).

A Field Guide to Geology: Eastern North America (Peterson Field Guides), by David C. Roberts (Houghton Mifflin, 1996). This guide covers the main highways of eastern North America with brief descriptions of the geology they traverse. Maps and diagrams tend toward the schematic but are usually adequate. The photo section is useful.

Fire, Faults & Floods: A Road and Trail Guide Exploring the Origins of the Columbia River, by Marge and Ted Mueller (University of Idaho Press, 1997). The book has a section on exploring the Channeled Scablands.

Fire Mountains of the West: The Cascade and Mono Lake Volcanoes, by Stephen L. Harris (Mountain Press, 1988). For lay readers, this is a detailed guide to volcanoes produced by the melting of the Pacific plate as it subducts under the western edge of North America.

Flood Basalts and Glacier Floods: Roadside Geology of Parts of Walla Walla, Franklin, and Columbia Counties, Washington, by Robert J. Carson and Kevin R. Pogue (Washington State Department of Natural Resources, 1996). The book covers the Channeled Scablands and Columbia River flood basalts.

Four Corners: History, Land, and People of the Desert Southwest, by Kenneth A. Brown (HarperPerennial, 1996). The book focuses on the region where Arizona, New Mexico, Colorado, and Utah meet.

Geologic History of Utah, by Lehi V. Hintze (Brigham Young University, 1988).

The Geologic Story of the National Parks and Monuments, by David V. Harris and Eugene P. Kiver (John Wiley & Sons, 1985). This useful volume covers all of the major parks and monuments with write-ups averaging several pages each and numerous photos and diagrams.

Geology of the Grand Canyon, by R. S. Babcock et al (Museum of Northern Arizona, 1974). This detailed compilation includes 12 authors writing chapters on the geology of various areas of the canyon and its history.

Geology of the Sierra Nevada, by Mary Hill (University of California Press, 1975). This small paperback gives a detailed look at Yosemite Valley and suggestions for touring in and around it.

Geology Underfoot in Death Valley and Owens Valley, by Robert P. Sharp and Allen F. Glazner (Mountain Press, 1999).

The Incomparable Valley: A Geologic Interpretation of the Yosemite, by François Matthes (University of California Press, 1950). While some of the geology is out of date, the book is a marvelous read. Besides, how many other geology books are illustrated largely with Ansel Adams photos?

An Introduction to Grand Canyon Geology, by Michael Collier (Grand Canyon Natural History Association, 1980). The canyon's geological history is written for the ordinary reader and illustrated by Collier's fine photos.

Roadside Geology of Alaska, by Cathy Connor and Daniel O'Haire (Mountain Press, 1988).

Roadside Geology of Arizona, by Halka Chronic (Mountain Press, 1983).

Roadside Geology of Hawaii, by Richard W. Hazlett and Donald W. Hyndman (Mountain Press, 1996).

Roadside Geology of Mount St. Helens National Volcanic Monument and Vicinity, by Patrick T. Pringle (Washington State Department of Natural Resources, 1993). Though not part of the *Roadside Geology* series from Mountain Press, this is a highly readable and well-illustrated guide to the volcano and its surroundings.

Roadside Geology of New Mexico, by Halka Chronic (Mountain Press, 1987).

Roadside Geology of Northern and Central California, by David Alt and Donald Hyndman (Mountain Press, 1999).

Roadside Geology of South Dakota, by John Paul Gries (Mountain Press, 1996).

Roadside Geology of Virginia, by Keith Frye (Mountain Press, 1986).

Roadside Geology of the Yellowstone Country, by William J. Fritz (Mountain Press, 1985).

A Streetcar to Subduction and Other Plate Tectonic Trips by Public Transport in San Francisco, by Clyde Wahrhaftig (American Geophysical Union, 1984).

MAGAZINES

Since the demise of *Earth* magazine in 1998 (look for back issues at your library), the main general-interest periodicals on geological subjects, *Rocks & Minerals* and *Rock & Gem*, are aimed at mineral collectors. Geology-oriented features appear often in general science periodicals such as *Discover* and *Scientific American*, and the U.K.-based *New Scientist*. The weekly *Science News* often has highly readable short reports and articles on new findings in the geological sciences; it's an excellent way to stay up to date on current research. For children, see *Odyssey* and *Scientific American Explorations*.

For those with some degree of expertise, *Science* and *Nature* publish significant scientific papers, usually accompanied by commentary that is somewhat less technical. Visiting a university library will turn up many publications for professional geologists.

Discover
111 Fifth Avenue, New York, NY 10011; tel: 800-829-9132; www.discover.com.

Nature
4 Crinan Street, London N1 9XW, England; tel: 44-171-833-4000; www.nature.com.

New Scientist
151 Wardour Street, London W1V 4BN, England; tel: 44-171-331-2701; www.newscientist.com.

Odyssey
30 Grove Street, Suite C, Peterborough, NH 03458; tel: 603-924-7209; www.odysseymagazine.com.

Rock & Gem
4880 Market Steet, Ventura, CA 93003-7783; tel: 805-644-3824; www.rockhounds.com.

Rocks & Minerals
5341 Thrasher Drive, Cincinnati, OH 45247; tel: 513-547-7142; www.mineralart.com/rocks_and_minerals.

Science
1200 New York Avenue, N.W., Washington, DC 20005; tel: 202-326-6501; www.sciencemag.org.

Science News
1719 N Street, N.W., Washington, DC 20036; tel: 202-785-2255; www.sciencenews.org.

Scientific American
415 Madison Avenue, New York, NY, 10017; tel: 800-333-1199; www.sciam.org.

Scientific American Explorations
179 South Street, 6th Floor, Boston, MA 02111; tel: 617-338-8231; www.explorations.org.

GEOLOGY ON THE INTERNET

These websites, in addition to those affiliated with organizations listed in this directory, offer information of use to those interested in the earth sciences. Most sites provide links to other geology-related websites.

www.aqd.nps.gov/grd/parks/byname.htm
Useful National Park Service guide to the geology of the parks and monuments.

geology.usgs.gov/gip.html
Site provides online U.S. Geological Survey publications for the general reader.

quake.wr.usgs.gov
The earthquake information site run by the U.S. Geological Survey has links for details of recent quakes and hazards in North America and around the world.

www.geophys.washington.edu/seismosurfing.html
The University of Washington's "Seismo-Surfing" page provides links to more than 100 earthquake research labs and departments around the world.

volcano.und.nodak.edu
VolcanoWorld at the University of North Dakota is aimed primarily at grades K-12, making it ideal as a beginner's resource.

vulcan.wr.usgs.gov
The Cascades Volcano Observatory, run by the U.S. Geological Survey, has loads of information about the volcanoes of the Pacific Northwest, including Mount St. Helens and Mount Rainier, their eruptions and hazards.

www.fs.fed.us/gpnf/mshnvm/volcanocam/
This website has a live camera pointed at the crater of Mount St. Helens from the Johnston Ridge Visitor Center; the picture is updated every few minutes, but keep in mind that clouds may at times obscure the mountain.

www.avo.alaska.edu
The Alaska Volcano Observatory keeps tabs on the state's active volcanoes.

hvo.wr.usgs.gov
The Hawaii Volcano Observatory sits on the edge of Kilauea's active caldera.

www.volcano.si.edu/gvp
Run by the Smithsonian Institution, the Global Volcano Network is the place to start researching volcanoes around the world.

www.rockhounds.com/rock-shop/linklist.html
This site, hosted by a rock shop, lists lots of links for rock and mineral collectors.

ORGANIZATIONS

Most national organizations concerned with geology and the earth sciences are for professionals. But many of the organizations listed below have publications for the general public and those interested in geological exploring.

American Geological Institute
4220 King Street, Alexandria, VA 22302-1502; tel: 703-379-2480; www.agi-web.org.

American Geophysical Union
2000 Florida Avenue, N.W., Washington, DC 20009-1277; tel: 202-462-6900; www.agu.org.

Bureau of Land Management
1849 C Street, N.W., Washington, DC 20240; tel: 202-452-5125; www.blm.gov.

Geological Society of America
3300 Penrose Place, Boulder, CO 80301-9140; tel: 303-447-2020; www.geosociety.org.

Mineralogical Society of America
1015 18th Street, N.W., Suite 601, Washington, DC 20036-5274; tel: 202-775-4344; www.minsocam.org.

National Park Service
Office of Public Inquiries, P.O. Box 37127, Washington, D.C. 20013; tel: 202-208-4747; www.nps.gov/parks.html.

United States Forest Service
201 14th Street, S.W. at Independence Avenue, S.W., Washington, D.C. 20024; tel: 202-205-1760; www.fs.fed.us.

United States Geological Survey
12201 Sunrise Valley Drive, Reston, VA 20192; tel: 703-648-4000; www.usgs.gov.

TOURIST INFORMATION

Alaska Tourism
P.O. Box 110801, Juneau, AK 99811-0801; tel: 907-465-2010.

Arizona Office of Tourism
1100 West Washington Street, Phoenix, AZ 85007; tel: 800-842-8257 or 602-542-8687.

Arkansas Tourism
1 Capital Mall, Little Rock, AR 72201; tel: 501-682-1088.

California State Division of Tourism
801 K Street, Suite 1600, Sacramento, CA 95814; tel: 800-462-2543 or 916-322-2881.

Hawaii Visitors and Convention Bureau
2270 Kalakaua Avenue, Suite 801, Honolulu, HI 96813; tel: 800-464-2924 or 808-923-1811.

Kentucky Tourism Council
1100 127 South, Building C, Frankfort, KY 40601; tel: 502-223-8687.

Massachusetts Office of Travel and Tourism
100 Cambridge Street, 13th Floor, Boston, MA 02202; tel: 800-447-6277 or 617-727-3201.

Missouri Division of Tourism
P.O. Box 1055, Jefferson City, MO 65102; tel: 800-877-1234 or 573-751-4133.

New Mexico Tourism
Lamy Building, 491 Old Santa Fe Trail, Santa Fe, NM 87503; tel: 800-545-2040 or 505-827-7400.

New York Tourism
1 Commerce Plaza, Albany, NY 12245; tel: 800-225-5697 or 518-474-4116.

South Dakota Department of Tourism
711 Wells Avenue, Pierre, SD 57501; tel: 605-773-3301.

Utah Travel Council
Council Hall, Capitol Hill, Salt Lake City, UT 84114; tel: 800-200-1160 or 801-538-1030.

Virginia Tourism
901 East Byrd, 19th Floor, Richmond, VA 23219; tel: 804-786-4484.

Washington State Tourism
P.O. Box 42500, Olympia, WA 98504-2500; tel: 800-544-1800 or 360-586-2088.

Wyoming Division of Tourism
I-25 at College Drive, Cheyenne, WY 82002; tel: 800-225-5996 or 307-777-7777.

PHOTO AND ILLUSTRATION CREDITS

AKG London 173

Bill Bachmann/Photophile 178

James Blank/Photophile 47T

Peter and Ann Bosted/Tom Stack & Associates 27B, 38, 142

Matt Bradley 81M

Dominique Braud/Tom Stack & Associates 89B

Dugald Bremner 129T, 139

Barbara Brundege/Ken Graham Agency 197B

Jonathan Chester/Extreme Images 33

Luigi Ciuffetelli 8T

Michael Collier 22, 35T, 41T, 95T, 121T, 138B, 141T, 172T, 181B, 201B

Patrick Cone 107T, 133TR, 181T

Betty Crowell 40T, 41B

Richard Cummins 34, 105T, 185M

Kent and Donna Dannen 101B, 102, 167L

Kevin Downey 14-15, 19, 68, 71T

John Drew/ImageArtist 10-11, 36 (Sandstone), 129B, 136B, 155B

John Elk III 74, 81T, 82, 85T, 85B, 89T, 172B, 174, 188, 189T, 193T

John Elkins/Travel Stock 205B

Jeff Foott/Jeff Foott Productions 5T, 8L, 12-13, 28-29B, 36 (Gneiss), 63, 89M, 90, 93T, 94, 95B, 96T, 96B, 98T, 98B, 106T, 132B, 141M, 152L, 154T, 157B, 160, 162, 163T, 164T, 164B, 165, 167R, 199B, 210-211

Jeff Foott/Tom Stack & Associates 18T, 101ML

Winston Fraser 51M, 51B

John Gerlach/Tom Stack & Associates 169B

Jon Gnass/Gnass Photo Images 177T, 193M

François Gohier 25T, 25B, 26T, 37 (Quartzite), 40B, 115T, 119B, 127T, 133TL, 170, 175

Manfred Gottschalk/Tom Stack & Associates 129M

J. Lotter Gurling/Tom Stack & Associates 182L

Thomas Hallstein/Outsight 18B, 169M, 177M

Bill Hatcher 1, 138T

Harvard University Archives 57

Christian Heeb/Gnass Photo Images 32T, 137T, 177B, 201T, back cover top

Kim Heacox/Ken Graham Agency 196, 199T

Paul Horsted 87T

George H. H. Huey 5B, 32B (upper right), 37 (Basalt), 105B, 109B, 117B, 125B, 134B, 141B, 144L, 149M, 149B, 204, 205T, 206-207B

John Hyde/Alaska Stock Images 194

Bruce Jackson/Gnass Photo Images 159M, 166

Gavriel Jecan 108

David Job/Ken Graham Agency 197T

Byron Jorjorian 81B

Wolfgang Kaehler 48T, 97T, 180, 186, 191

Lewis Kemper 99, 111T, 152M, 152R, 155T, 156T, 156B, 157T

Breck P. Kent 36 (Peridotite, Granite, Conglomerate, Schist), 37 (Rhyolite, Shale, Limestone, Marble)

Thomas Kitchin/Tom Stack & Associates 47B, 201M

G. Brad Lewis/Photo Resource Hawaii 202

Robert Llewellyn 60

J. Lotter/Tom Stack & Associates 185B

Alan Majchrowicz 29T, 185T, 189B, 190B

Stephen Matera 42-43

Buddy Mays/Travel Stock 9T, 59M, 76, 77B, 78M, 78B, 79, 149T, 183

Joe McDonald/Tom Stack & Associates 71B

Steve Mulligan 86-87B

William Neill 6-7, 56T, 70, 107B

Mark Newman/Tom Stack & Associates 206T

Richard T. Nowitz 44, 48B, 49, 86T

Laurence Parent 145, 147, 159T

Brian Parker/Tom Stack & Associates 144R, 209M

Allen Prier/Ken Graham Agency 9B

Louie Psihoyos 127B

Walt Puciata/Gnass Photo Images 121B

Milton Rand/Tom Stack & Associates 182R

Paul Rezendes 4, 27T, 35B, 51T, 52, 54-55B, 56B, 59T, 62, 64T, 65, 67B

Sam Roberts 163B

Stephen J. Shaluta, Jr. 73B

Smithsonian Institution 134T

Scott Spiker 116, 119T

Tom Stack/Tom Stack & Associates 207T

Spencer Swenger/Tom Stack & Associates 136T, 150

Loren Taft/Ken Graham Agency 24T

Tom Till 2-3, 28T, 30, 32B (upper left), 32B (lower left), 32B (lower right), 55T, 59B, 64B, 67M, 73T, 73M, 77T, 78T, 92B, 97B, 101T, 101MR, 106B, 109T, 111B, 112, 115B, 117T, 121M, 122, 125T, 126T, 135, 153, 154B, 198

Stephen Trimble 16, 20-21, 111M, 130, 137B, 146L, 159B, 193B, 209B

Larry Ulrich front cover

University of Chicago Library 190T

Greg Vaughn/Tom Stack & Associates 169T, 209T

Wiley/Wales 118, back cover bottom

George Wuerthner 67T, 93B

Design and layout by Mary Kay Garttmeier

All maps and illustrations by Karen Minot except 146R, illustration by Karen Minot based on a drawing by Deborah Reade

Index by Elizabeth Cook

T-top, B-bottom, R-right, L-left, M-middle

INDEX

Note: page numbers in italics refer to illustrations